UNDER THE MAP OF GERMANY

UNDER THE MAP OF GERMANY

Nationalism and propaganda 1918–1945

Guntram Henrik Herb

London and New York

First published 1997
by Routledge
11 New Fetter Lane, London EC4P 4EE

Simultaneously published in the USA and Canada
by Routledge
29 West 35th Street, New York, NY 10001

Typeset in Garamond by
J&L Composition Ltd, Filey, North Yorkshire
Printed and bound in Great Britain by
Biddles Ltd, Guildford and King's Lynn

British Library Cataloging in Publication Data
A catalogue record for this book is available from the British Library

Library of Congress Cataloging in Publication Data
Herb, Guntram Henrik
Under the Map of Germany: Nationalism and Propaganda 1918–1945 /
Guntram Henrik Herb.
p.   cm.
Includes bibliographical references and index.
1. Germany – Historical geography.   2. Germany – Politics and
government – 1918–1933.   3. Germany – Politics and
government – 1933–1945.   4. Nationalism – Germany.   5. Propaganda,
German.   6. Geopolitics – Germany – History.   7. Germany – Relations –
Europe.   8. Europe – Relations – Germany.   I. Title
DD21.H47   1997
911′.43′09041 – dc20   96-7437
ISBN 0–415–12749–1

CONTENTS

CONTENTS

FIGURES

vii

Note: The maps reproduced in this volume are historical documents and the author and publisher have made every effort to obtain the best quality copies available.

ACKNOWLEDGEMENTS

Much of this book is based on research originally carried out for my Ph.D. thesis at the University of Wisconsin. Since the completion of the dissertation in 1993, I had time to get enough critical distance to rework the material and prose, to place the study in a larger context, and to incorporate recent works. Comments by two anonymous reviewers and Tristan Palmer at Routledge have been of great help in this process. Over the course of the more than eight years since I first started work on the topic, I have accumulated a tremendous amount of debt to individuals and organizations. First and foremost, I want to thank Mark Bassin, who has been a mentor and an inspiration, and David Woodward, whom I deeply appreciate for his unwavering support and encouragement.

Many other individuals contributed to the successful completion of this work through commenting on drafts, assisting with references and resource materials, and engaging me in stimulating discussions: Malcolm Cutchin, Matthew Edney, Anne Godlewska, Theodore Hamerow, the late Brian Harley, Joel Harrington, the late Richard Hartshorne, Dave Kaplan, Kevin Kaufman, Andrew Kirby, Robert Koehl, Klaus Kost, Marjorie Lamberti, M. Lazar, Jude Leimer, Joan Nogué-Font, Mechthild Roessler, Gerhard Sandner, and Frank Wrobel.

I would also like to express my general gratitude for all the assistance I received at the various archives and institutes I visited. Special mention should be made of Dr Hofmann at the Bundesarchiv in Koblenz, Verena Kleinschmidt and Günther Bouché at the Georg Westermann Werkarchiv in Braunschweig, who were extraordinarily helpful. My research received generous funding from the National Science Foundation (Grant No. SES 89-10702), the University of Wisconsin Alumni Research Foundation, and the American Geographical Society in Milwaukee, for which I remain extremely grateful.

Above all, it is my family who deserve my thanks. I dedicate this book to them. My parents and brothers made immeasurable contributions, especially my father, Martin Herb. My children, Henrik and Natalie, provided a welcome source of distraction and inspired me to tackle new challenges.

ACKNOWLEDGEMENTS

The main driving force behind the completion of the project was my wife, Patricia. To her I am particularly grateful for her companionship, patience, and love.

Note: Every effort has been made to contact all copyright holders. However, if any have been inadvertently omitted the copyright holder should contact the publisher who will be pleased to make the necessary arrangements at the first opportunity.

ABBREVIATIONS

AA	Politisches Archiv des Auswärtigen Amtes, Bonn
AGS	American Geographical Society Collection, Milwaukee
BA	Bundesarchiv, Koblenz
BDC	Berlin Document Center
BDO	*Bund Deutscher Osten*
BfPB	Archiv der Bundeszentrale für Politische Bildung, Bonn
DAI	*Deutsches Ausland-Institut*, Stuttgart
DKG	*Deutsche Kartographische Gesellschaft*
GEI	Georg-Eckert-Institut für Internationale Schulbuchforschung, Braunschweig
NSDAP	*Nationalsozialistische Deutsche Arbeiterpartei*
NSLB	*Nationalsozialistischer Lehrerbund*
NODFG	*Nord- und Ostdeutsche Forschungsgemeinschaft*
Puste	*Publikationsstelle-Berlin*
PRO	Public Record Office, Kew/London
RfH	*Reichszentrale für Heimatdienst*
Stiftung	*Stiftung für deutsche Volks- und Kulturbodenforschung*, Leipzig
VDA	*Verein/Volksbund für das Deutschtum im Ausland*
Vomi	*Volksdeutsche Mittelstelle*
WWA	Georg Westermann Werkarchiv, Braunschweig

INTRODUCTION

What is the fatherland of the German? As far as the German language is heard.

(Ernst Moritz Arndt 1813)[1]

We demand the unification of all Germans in one Greater Germany on the basis of the right of national self-determination.

(1. Programmpunkt der NSDAP)[2]

The goal of national self-determination has always played a central role in German history. It also has given German nationalism an expansionist tendency because the German state rarely included the entire German nation. The most extreme manifestation of this expansionist tendency occurred in Nazi Germany, which justified most of its annexations and early conquests with the right of ethnic German self-determination. The vast majority of Germans believed in the rightfulness of these territorial acquisitions; it considered the Sudetenland, Austria, the Memel territory, and Alsace-Lorraine as well as parts of Poland to be unquestionably German national territory, despite the sizable and at times even majority presence of other nationalities. This view of a "Greater Germany" is still shared by a considerable number of Germans today. In 1989, the New Right German party *Die Republikaner* disseminated a map which claimed virtually the same areas for Germany in the name of national self-determination as the early Nazi conquests.[3]

East and West Germany have been unified since the publication of the *Republikaner* map and the German government has signed several treaties, such as the German–Polish treaties on boundaries (November 1990) and on friendship and cooperation (June 1991), but the issue of German national territory is still not laid to rest. Associations for Germans expelled after the war from Eastern Europe (*Vertriebenenverbände*) continue to stress German rights to areas east of the current German state borders and New Right ideology has been adopted by members of mainstream parties, such as the Christian Democrats. An important element is the increase in revisionist historical scholarship, which led to the "historians' dispute"

1

(*Historikerstreit*) in the late 1980s.[4] By equating Nazi atrocities with those committed under Stalin and denying full German responsibility for the Second World War, these neo-conservative historians not only challenge the singular character of the holocaust, but also imbue Nazi Germany with legal respectability. This encourages demands for German self-determination in "Greater Germany."[5]

How and when did this consensus originate about the extent of the greater German nation or "Greater Germany"? I argue that a consensus was neither brought about by skillful Nazi propaganda nor evolved slowly since the beginnings of German nationalism in the mid-nineteenth century. Rather, it was created during the Weimar Republic through a massive map campaign. The trigger event for this campaign was the Treaty of Versailles, which was wholeheartedly rejected by Germans as unjust. Their argument was that Germany had been denied national self-determination, despite the international recognition of this principle at the end of the First World War. Yet, at the time, there was in Germany neither a clear consensus about the territorial extent of German self-determination, nor sufficient data to make a strong case for it.

The need for convincing and universally accepted German territorial demands brought together two groups: *völkisch* activists advocating the supremacy of Germans, and geographers – experts on territorial aspects of the German nation.[6] Especially after encountering several foreign maps making effective arguments for territorial concessions from Germany, geographers and *völkisch* nationalists realized that maps were the key to their revisionist aspirations. Imbued with scientific respectability, cartographic representations were the only means to convey a clear image of the boundaries of German national territory. As a result, the discourse of German self-determination became thoroughly cartographic. Maps were completely subordinated to the national cause and a comprehensive cartographic attack was launched to define and popularize the "just" extent of the German nation. The well-defined objective of this attempt to change public opinion, and the deliberate and organized manner in which it was pursued, characterizes this endeavor as propaganda (Jowett and O'Donnell 1986: 15–16; Welch 1993: 5).

The successful propaganda map campaign by *völkisch* groups and geographers in the Weimar period dispels the myth that Nazi Germany was one of the most masterful producers of cartographic propaganda. This view, which is reiterated throughout the literature (e.g. Tyner 1974 and 1982; Prestwick 1978; and Pickles 1992), can be traced back to the Second World War. American authors from that period, such as Speier (1941), Strausz-Hupé (1942), Quam (1943) and others alerted the public to the effectiveness of German propaganda maps. Although these wartime works were able to identify general principles of cartographic manipulation – for example, the use of disproportionate symbols in comparisons of minority population – they nevertheless presented a distorted picture. They assumed that Nazi propaganda mapping was a monolithic enterprise centered on an *Institut für Geopolitik* in Munich. Clearly, American authors from this period were unable to obtain first-hand information and their view on German propaganda mapping was influenced by the then current notion that the German school of *Geopolitik* provided Hitler's blueprint for expansion.

Recent studies by Jacobsen (1979) and Bassin (1987b) have revealed that the wartime notion of a central role of German *Geopolitik* in National Socialism is inaccurate. *Geopolitik* was not a unified school of thought – the *Institut für Geopolitik* never existed – and the two theories had fundamental differences. The same is true for the production of propaganda maps. I have shown elsewhere (Herb 1989) that persuasive mapping by followers of *Geopolitik* cannot be equated with National Socialism. Here I go one step further by proving that the cartographic propaganda methods customarily believed to be Nazi achievements were already developed in the Weimar Republic and by a variety of individuals and institutions.

The links made in this book among maps, cultural context, and political concepts, such as national territorial identity, fulfill a demand that has been increasingly voiced in the literature (e.g. Harley and Woodward 1987: 502–9; Harley 1988a and 1992; Pickles 1992). Most works in cartography and political geography still narrowly focus on aspects of design elements of maps – such as color, scale, and projection – and simply assume that a monolithic government structure is behind their production (e.g. Speier 1941; Quam 1943; Tyner 1974 and 1982; and Prestwick 1978). Wilkinson's (1951) analysis of the mapping of Macedonian ethnographic territory looks at the political influence of maps. However, he overemphasizes cartographic accuracy and neglects the institutional structures behind their dissemination and production. Examinations of the propaganda process in general (Jowett and O'Donnell 1986) or case studies (e.g. Sywottek 1976; Wippermann 1976; Humbel 1977; Paul 1990; Welch 1993) deal with institutions and different media, but in most cases exclude maps.

My examination of nationalist activities spanning from the Weimar Republic to the Third Reich reveals how crucial the involvement of private groups and particularly scientific disciplines, such as geography and cartography, was in the creation of a consensus about German national territory. Private initiatives and voluntary subordination of research under nationalist goals were more important than governmental activities, not just in democratic Weimar, but even in the dictatorship of the Third Reich. These findings complement studies which look at the way scientific concepts are related to state propaganda (Heske 1986; Bassin 1987a), at the involvement of geographical institutes and geographers in National Socialism (Roessler 1990; *Geographie und Nationalsozialismus* 1989), and at the larger social context of Nazi propaganda (Welch 1993).

The continual activities by the alliance of scholars and *völkisch* groups

3

from the end of the First World War to the Nazi period has other important repercussions. First, it shows that much of the expansionism of the Nazi state had been made palatable and convincing to the public as early as the 1920s. The groundwork for National Socialist conceptions of what was "rightfully" German territory had been laid then. These aspects were omitted in the detailed work by Rich (1973 and 1974). Second, the choice of the Nazis to collaborate with the nationalist alliance, rather than to transform it into a centralized organization, reveals the complexity of the Nazi power structure. This provides further empirical evidence for the argument in the literature that totalitarian theories about the National Socialist state are inadequate.[7] It also extends this argument to one of the last vestiges of the totalitarianism school: most of the literature on propaganda cartography.

On a more general level, this study gives insights into the territorial component of national self-determination. National self-determination is considered a just cause in international relations. Although not a formal part of International Law, it is one of the fundamental goals of the United Nations Charter (article 1.2) and customarily cited to support territorial claims. Since national groups intermix and there is no uniform concept of what constitutes a nation, conflicts about claims are inevitable. A recent example is the war in Bosnia, in which Croats, Muslims, and Serbs were fighting for control over their overlapping national territories.

Another question touched upon in this work, whether it is possible to delimit national territories in a "just" way, also has not been addressed in the literature, except for German works from the study period – which are clearly biased. There is an enormous literature dealing with nationalism and territory, but the actual delimitation of national territories is side-stepped. Publications either focus on general aspects of territoriality as a means of exercising power (Sack 1986; Gottmann 1973), on the general relationship between territory and nationalism (J. Anderson 1988), on the theoretical foundations of nationalism or the concept of nation (Breuilly 1993; B. Anderson 1991; Mayer 1987), on the ideas behind the construction of national space (Williams and Smith 1983), or on the historical development of nations (Hobsbawm 1990; Seton-Watson 1977).

The book is organized into three parts. The first part (Chapters 1–2) provides the background necessary for the detailed examination of the consensus creation in the Weimar Republic. Apart from clarifying theoretical and terminological issues, it examines the transformation of national self-determination at the end of the First World War. Viewed through the lens of cartographic representations, this includes the specific German case as well as the larger context of the Allied peace preparations. Part two (Chapters 3–7) analyzes the new emphasis on the territorial limits of the German nation during the Weimar Republic. It traces the effort of a nationalist alliance of *völkisch* groups and geographers to convince the

German people of their right to a Greater Germany. Of particular importance are the organizational structures behind this mapping effort and the role of scholars in justifying and maximizing claims. The final part (Chapters 8–9) shows how the activities of the nationalist alliance continued during the Third Reich. The guiding issues are the Nazi regime's attitudes toward German self-determination and maps in general.

1

NATIONALISM, TERRITORY, MAPS, AND PROPAGANDA

Where political and national boundaries coincide, society ceases to advance . . .

(Lord Acton 1907)[8]

We are fooled by propaganda chiefly because we don't recognize it when we see it.

(Institute for Propaganda Analysis 1937)[9]

The historical roots of the idea of "nation" can be traced back at least to the Middle Ages, but it only became a decisive political force at the end of the eighteenth century with the emergence of the doctrine of nationalism and the notion of popular sovereignty (Seton-Watson 1977: 6–8). There is no universal definition for nation, but all nations make reference to a unique and separate identity and aspire to achieve self-determination.[10]

Despite the rhetoric of nationalists, nations are not organic entities with a common ancestral descent. The separate identity has to be understood as something that is collectively self-defined. Nations are "imagined communities" (B. Anderson 1991) whose consciousness of belonging together is founded on a "national myth" (Connor 1992). To preserve this feeling of being a unique group, nations yearn to be independent, which inevitably leads to the demand for a separate state. Only states are recognized by the international community as sovereign entities and thus have undisputed authority.[11]

Territory plays a key role in nationalism. National identity cannot be separated from links to a specific territory which provides the only tangible basis for the national myth. The homeland is the repository of national history. It is the place where the nation has its roots. The mountains of the national landscape are sacred, its rivers carry the national soul, its soil is soaked with the blood of national heroes (Williams and Smith 1983: 509). Territory is also the only tangible basis on which to achieve the desired national self-determination because sovereign political power is exercised by states over a specific and clearly demarcated territory (Gottmann 1973).[12]

6

The territorial or spatial nature of nationalism encourages the use of cartographic depictions. Maps as "graphic representations that facilitate a spatial understanding of things, concepts, conditions, processes, or events in the human world" (Harley and Woodward 1987: xvi) are ideally suited to further the cause of nationalist movements. They can be used to demarcate the limits of a nation and to communicate these limits to others to create a consensus. Clear territorial limits are needed to fulfill the goal of nations to be exclusionary communities: to distinguish "us" from "them."

Defining the limits of a nation without the help of graphic images is difficult, if not impossible. Even if references are made to natural landscape features, the message is rarely clear or unambiguous. For example, the first verse in the German national anthem (which is illegal today) conveys a general idea of the parameters of the German nation. It mentions rivers and straits in the four cardinal directions: "*Von der Maas bis an die Memel, von der Etsch bis an den Belt.*" But these geographical landmarks are not common knowledge and they still leave out considerable areas in the northwest, the southwest, and the southeast. In the case of the French nation, the Pyrenees are often used as a southern limit. But where in this mountain range is the national divide? In the northern foothills, along the ridge line, or somewhere on the southern slopes? And what if there are no clear natural features as in the case of the Eastern European plains? Only cartographic representations can communicate a clear image of the boundaries of national territory.

Maps as means of communication here does not presume that maps are mirrors of reality. The communication model used by many cartographers, in which maps are objective tools for transmitting information, is flawed (Pickles 1992: 194–7). It presumes that cartography is a neutral science which constantly seeks to make representations more and more accurate, to bring them more and more in line with reality. However, the production of maps cannot be separated from the societal context in which it occurs.[13]

Even the seemingly most accurate map is still a transformed and thus an interpreted picture of reality. A case in point is official topographic maps which are generally considered scientific and "objective" documents. How can we explain the persistent inclusion of battlefields and other monuments deemed important to the nation, and the steadfast exclusion of toxic waste sites, abandoned public housing units, and other structures potentially threatening to the cohesion of our national community (Monmonier 1991: 122)? The information that is represented had to be collected, classified, and encoded; a process that is structured by social norms and values, regardless of whether the cartographer is aware of it or not.[14]

Like all knowledge, maps are expressions of power; they are inherently rhetorical. Thus, to understand the role of maps in the construction of a national territorial identity, maps have to be deconstructed and analyzed in their "layers of textuality": the cartographic image itself, the material it

7

accompanies, and the larger social context (Pickles 1992: 219). National identity is an artificial construct which is conceptualized and disseminated through social discourse. Maps have to be viewed as part of this discourse, as simply another text.[15]

Even though maps are rhetorical texts and cartography is "an art of persuasive communication" (Harley 1989: 11), this does not mean that maps or cartography in general should be considered propaganda. There is much confusion about the terms rhetoric, persuasion, and propaganda in the literature and they are often used interchangeably. Propaganda is a much narrower concept; it can be characterized as a *"deliberate* and *systematic* attempt to shape perceptions, manipulate cognitions, and direct behavior"* (Jowett and O'Donnell 1986: 16; emphasis added). By contrast, persuasion or rhetoric do not attempt to restrict the flow of information and to conceal the overall purpose of the interaction; they are more neutral terms than propaganda (Jowett and O'Donnell 1986: 15–24). At the same time, propaganda should not be equated with lies and falsity or with complete changes in attitudes and beliefs; it often contains an element of truth and exaggerates existing trends (Welch 1983: 2 and 1993: 5).[16]

In Germany, national cartographic discourse did not involve propaganda in the true sense in the early years. The appearance of maps of the German nation up to the First World War was sporadic and haphazard. Maps tended to be more individual responses to major events than products of deliberate and systematic campaigns. They reflect the absence of a clear concept about the limits of the German nation before 1871.

EARLY REPRESENTATIONS OF GERMAN NATIONAL TERRITORY

When a German national consciousness began to emerge at the turn of the eighteenth to the nineteenth century, it was guided by the ideas of Herder, which looked at German language and culture as a common national bond. However, there was no unified stance on how the German nation should be incorporated into a state structure, nor was it politically viable until 1848. Initial calls for national unification, such as Johann Gottlieb Fichte's famous *Addresses to the German Nation* in the winter of 1807/8 and Ernst Moritz Arndt's German songs (*Lieder für Teutsche*) apparently did not enjoy widespread support (Langewiesche 1992b: 353–4). A major problem of German self-determination was that the area inhabited by Germans was not centralized spatially. In Eastern Europe German settlements were widely and often thinly dispersed, which made it difficult to arrive at a clearly defined and generally recognized territorial base (Bassin 1987a: 475). The only agreement was that foreign rule by Napoleon and German particularism, that is, the division of the German area into individual principalities, had to be overcome (Schieder 1991: 146–7).

The first maps of the German nation did not appear until the 1840s, shortly before the first national revolution in 1848 led to a German national assembly at St Paul's Church in Frankfurt. As is customary for most maps of national territory, they used language as an indicator. Apart from political upheavals in Germany, these cartographic representations appear to have been stimulated by concerns about large-scale emigrations in the 1830s and 1840s (Fittbogen 1930: 49). This interest in the overall distribution of the German language was complemented by a local research interest into areas where Germans faced competition from other national groups, such as the emerging Italian nation (Fittbogen 1927: 26–7, 74).

The language map of Germany from 1844 by the scholar and politician Karl Bernhardi is the first recorded attempt at an overview. It was presented to the philologists and teachers who assembled in Kassel in October 1843 "for examination and sponsorship" and still had grave flaws, such as labeling Klagenfurt as Slavic (Isbert 1937a: 491–2).[17] Work on the map had begun in 1834 with the support of fourteen German historical societies and under the leadership of Baron von Hormayer (Weidenfeller 1976: 63). It was followed in 1848 by Heinrich Kiepert's map of the German nation (Isbert 1937a: 492).

There was a lot of discussion about the boundaries of the German nation in the national assembly in Frankfurt, but it was clear that the future German national state would respect the integrity of the individual principalities and only attempt to unify them in a federal system.[18] The main point of controversy was which of the rulers of the two largest German states, Prussia and Austria, would become the head of the German national state. The choice of the Austrian Habsburgs, known as the *großdeutsch* solution, included all German settlement areas but also a large number of minorities. By contrast, a preference for Prussian dominance along the lines of the German Customs Union, dubbed the *kleindeutsch* solution, meant the exclusion of the Germans in Austria. In the end the assembly opted for the *kleindeutsch* solution, but the revolution failed in 1849.

With the sovereignty of the individual German principalities restored, scholarly work on German national culture and its cartographic representation declined (Fittbogen 1930: 50–1). The second edition of Bernhardi's map, which had been prepared in collaboration with the Frankfurt physician Wilhelm Stricker in 1849, was indicative of this negative trend.[19] It was dedicated to the members of the national constitutional assembly "in *memory* of the lively discussions about the natural borders of the German Empire."[20] By stressing the importance of the map as a "memory" of the discussion, it looked backward rather than presenting the map as a program of what should be accomplished in the future.

The next forty years were not very conducive for the production of maps of the German nation. Nationalist activities were oppressed during the period of reaction which followed the failed revolution, and for the first

9

decade after the founding of the German Empire, national consciousness focused on the political territory of the new German state. Only one map was published in 1867, around the time when it was clear that Bismarck was striving to unify German territories under Prussian tutelage.[21] The motivation behind this map might have been to raise public awareness about the discrepancy between the achievements of Bismarck and the full extent of the German nation.[22]

When a German national state was finally established in 1871, it was not the result of a popular revolution, but the product of Bismarck's policy of unification from above. Despite the fact that it excluded the Germans in the Austro-Hungarian Empire – Bismarck had only realized a *kleindeutsch* solution – the establishment of the national state from above kept the loyalties to the state and the dynastic rulers intact (Fischer 1977: 13–14). The people identified with their fatherland, i.e. the German Empire, rather than with their fellow Germans abroad (Jaworski 1978: 371). The political boundaries of the new German national state respected the existing dynastic territorial units. This meant that areas with a clear majority of foreigners, such as the Polish districts in Prussia, were not excluded from the unification even though their separation from the Empire would have resulted in a more homogenous "national" German state. The national idea was also disregarded in cases where the German government made territorial gains: it was more interested in expanding its military base and economic might than in extending its control over all German people.

The case of Alsace-Lorraine illustrates this point. When Bismarck annexed these territories in 1871, he was not guided by the idea of national self-determination, but by strategic considerations. Owing to pressure from the German military to establish a strong military frontier against France, the final course of Germany's western boundary went well beyond the German-speaking regions on the left bank of the Rhine.[23]

Even though the German Empire of 1871 with its priorities on military security gave only secondary consideration to the German national idea, the success of finally achieving a unified German state gave German nationalism a decisive boost. In the 1880s a variety of nationalist organizations began to appear such as the *Flottenverein* and the *Alldeutscher Verband*.[24] The nationalist ideology expounded by them was complex and changed over time, but there were two main camps: a group which identified with the new state, and an ethnic national movement yearning to be united with fellow Germans outside the state borders. Both movements were expansionist. The key difference is the type of community they identified with most strongly.[25]

The state-centered nationalism looked at the population of the German Empire as the "nation", following Meineke's idea of a *Staatsnation* or political nation. It sought to make the Reich into a global imperial power through naval build-up and the acquisition of colonies. The ethnic nationalist movement, which was much less significant, felt first and foremost part of the brotherhood of all Germans.[26] Its primary focus was expansion in Europe to "redeem" all Germans who were at present excluded from the German state.[27] For some of the ethnic nationalists this meant the attempt to fulfill the *großdeutsch* ambitions which had been foiled by the foundation of the German Empire in 1871, for others it entailed a *völkisch* mission to conquer Europe and the world.[28]

Despite the invigorated nationalism of the 1880s, the academic community, especially geographers who were experts on maps, did not pay much attention to the mapping of German ethno-national territory.[29] One of the reasons is that up to the beginning of the First World War, most geographers were proponents of the idea of a political nation, which meant that they identified more with the population inside the existing German state than with all ethnic Germans (Faber 1982: 395–8). They believed that state boundaries should be based on geographical aspects (i.e. natural features) rather than ethnic considerations (e.g. Ratzel 1897). Much of the work evolved around defining the geographical basis of a German-dominated *Mitteleuropa* or Central Europe (Schultz 1987, 1989a, 1989b).

Another reason for a lack of interest in ethnic German territory is that geography had only been established as an academic discipline by the German government in the wake of the foundation of the German Empire in 1871 (Hudson 1977). In their search for official recognition, the vast majority of geographers strove to show their unwavering support and usefulness for the existing state (Sandner and Roessler 1994: 115–16). This made it politically inopportune to draw attention to the incomplete nature of national unification, which would have entailed a serious challenge to the legitimacy of the new German Empire.[30]

Up to the beginning of the First World War, the mapping of German national territory was a purely *völkisch* endeavor. The most active organization was the *Deutscher Schulverein*, founded in 1881 and later renamed *Verein für das Deutschtum im Ausland*,[31] which commissioned and disseminated two large-scale overview maps.[32] The other maps were products of the *völkisch* geographer Paul Langhans.[33] Langhans, who was infamous for his extreme nationalism or *Deutschtümelei*, worked for the Perthes publishing house in Gotha.[34]

With the encouragement of the *Alldeutscher Verband* Langhans also published an atlas of the distribution of Germans in 1900 (Meyer 1946: 182). It was entitled *Alldeutscher Atlas* and criticized abroad as being irredentist (Köhler 1987: 168).[35] When he took over the editorship of the respectable journal *Petermanns geographische Mitteilungen* in 1909, many foreign authors decided not to contribute to it any more (Köhler 1987: 165). Even his fellow German geographers, such as Hermann Wagner, were suspicious of his nationalism and questioned his scholarly integrity (Köhler 1987: 151).

11

A general characteristic of these maps by *völkisch* groups is the stress on the wide dispersion of the German language in Europe and all over the world. This served two purposes: 1) it gave evidence of the greatness of German culture by showing its far-reaching influence and 2) it revealed that German culture faced the danger of dissolving into foreign cultures on the outer fringes of the contiguous German-language area.[36]

The stimulus for *völkisch* activities seems to have come mainly from the Austro-Hungarian Empire, where relationships between Germans and other national groups were deteriorating,[37] and from Polish–German tensions in Prussia. Restrictive laws were passed by the German government on language use and other cultural expressions which alienated the Poles and resulted in a significant increase in Polish nationalism (Holborn 1969: 294). The growth of the Polish national movement in turn stimulated German nationalism because it evoked a fear of a threat to the German state from the Slavic East.

The belief that Germans faced an onslaught of Slavs from the East – a major component of *völkisch* ideology – explains why *völkisch* activists did not exaggerate the presence of Germans in these areas. They saw no need to establish claims because these areas were already part of the German or Austro-Hungarian Empire, but rather tried to show the dangers of not having a contiguous German settlement region. This intention came back later to haunt them: most of the eastern territories which Germany had to cede later in the Treaty of Versailles were depicted as Polish on maps made by *völkisch* organizations.[38]

In the early 1900s and especially after the outbreak of the First World War, which created an immediate sense of unity among Germans, *völkisch* ideas became more popular. The prevailing view of a political nation among geographers was slowly being challenged by a younger generation of geographers who taught in areas such as Austria, where conflicts between nationalities were a daily occurrence (Faber 1982: 398). They included Albrecht Penck in Vienna,[39] Joseph Partsch in Breslau, and Robert Sieger in Graz. However, these developments did not lead to the production of new maps until after the armistice negotiations revealed the importance of the right of national self-determination for the post-war settlement.[40]

During the war, right up to the armistice, most German geographers were swept away in the expansionist mood created by the war: they were not interested in national self-determination, which would have seriously limited their claims. Rather, they employed a variety of principles, such as easily defensible boundaries and the idea of natural regions, to maximize German demands.[41]

2

CARTOGRAPHY AND NATIONAL TERRITORY AT THE END OF THE FIRST WORLD WAR

THE FIRST WORLD WAR AND NATIONAL SELF-DETERMINATION

There can be no doubt as to the intentions of the Allied and Associated Powers to base the settlement of Europe on the principle of freeing oppressed peoples, and redrawing national boundaries as far as possible in accordance with the will of the peoples concerned.

(Temperley 1969 II: 262)

The concept of national self-determination acquired unprecedented importance at the end of the First World War (Knight 1984: 171). It became the cornerstone of the professed war effort of the Allied powers, France, Great Britain, and Russia. It enabled them to elevate the war above a purely national affair. They could rally the support of their people by declaring that this was a "just" war which was to give to the oppressed nationalities what was rightfully theirs.[42]

Yet, these war slogans and the true war aims during the first years of the war differed markedly. Various treaties and agreements among the Allied powers outlined territorial gains as compensation for the war effort which did not respect national self-determination.[43] It was only toward the end of the war that the Allies truly began to embrace the movements for national self-determination in Eastern Europe. Only then did it become clear that the struggle of the small nations in Eastern Europe for independence was of considerable help for the Allied cause (Mantoux 1952: 35). France's Prime Minister Georges Clemenceau admitted to this later:

Alas, we must have the courage to say that our programme, when we entered the War, was not one of liberation! . . . The surrender of Russia . . . changed the data of the problem, by grouping round us forces striving for national restoration, which had been incompatible with the presence of the Czar in our ranks. . . . We had started as allies of the Russian oppressors of Poland, with the Polish soldiers of Silesia and

13

Galicia fighting against us . . . our war of national defence was transformed by force of events into a war of liberation.

(Clemenceau 1930: 190–2)

The national independence movements not only helped the Allies in their
fight against Germany and Austria-Hungary, but also were seen as a buffer
to the spread of Bolshevism. The possibility of a buffer zone of small
independent states in Eastern Europe offered two benefits: it would
provide a *cordon sanitaire* against communist Russia and it would weaken
Germany.

Although this might appear surprising, the Soviet Union also advocated
national self-determination. However, it wanted the principle applied universally, that is, not only in Eastern Europe, but also in the rest of the
world. The intent was obvious, since universal national self-determination
would have severely disrupted the capitalist economies of the West by
making the colonies independent. In addition, by publishing the text of
the treaties to which Russia had been a signatory, such as the secret Treaty
of London (1915), the Soviet Union tried to induce political opposition in
the Allied countries by exposing the territorial "greed" of the capitalist
West.

The just cause of war, i.e. granting self-determination to oppressed
national groups in Eastern Europe, was also of central importance to the
war effort of the United States. How else could President Wilson justify to
the American people getting involved and sending American troops into
battle? Wilson truly believed in this moral mission. He had the utopian goal
of establishing a new world order which would ensure a lasting peace,
especially in war-ridden Europe. His idea was to pacify Europe by giving
each national group sovereignty over its own territory (Gelfand 1976: 141–
2). Wilson's peace policy was outlined in his famous Fourteen Points
Message of 8 January 1918. In points 9, 10, 11, 12, and 13, respectively,
he stressed self-determination as a solution to the main territorial problems
of Italy, Austria-Hungary, the Balkans, the Ottoman Empire, and Poland.

The Entente had affirmed the principle of national self-determination in
the declaration of its war aims already on 10 January 1917, but Wilson's
Fourteen Points were much more crucial: they became the basis for the
armistice of 11 November 1918 and the peace treaty which followed. As
the Allied and Associated Powers stated in a letter of 16 July 1919 to the
German Delegation:

> The Allied and Associated Powers are in complete accord with the
> German Delegation in their insistence that the basis for the negotiation
> of the Treaty of Peace is to be found in the correspondence which
> immediately preceded the signing of the Armistice on November 11,
> 1918. It was there agreed that the Treaty of Peace should be based
> upon the Fourteen Points of President Wilson's address of January 8,

1918, as they were modified in the Allies' memorandum included in the President's note of November 5, 1918, and upon the principles of settlement enunciated by President Wilson in his later addresses, and particularly in his address of September 27, 1918.

(cited after Temperley 1969 II: 249).

Thereby, the principle of national self-determination was officially recognized for the first time by all major world powers.[44]

WILSON'S FOURTEEN POINTS AND THE IMPORTANCE OF MAPS

Wilson's Fourteen Points elevated not only national self-determination to central importance in political discourse, but also maps. The US president had developed his points with the help of the Inquiry, a commission which had been founded a few months earlier, in September 1917. Housed at the American Geographical Society of New York, the center for cartographic work in the United States (Gelfand 1976: 274), the Inquiry mostly consisted of scholars, such as historians, economists, political scientists, and geographers. Maps were extremely important for the work of the Inquiry, and it was realized from the beginning that a "thoroughgoing map program" needed to be established (Gelfand 1976: 350). Initially, cartographic depictions even helped Wilson in formulating his Fourteen Points. Just prior to giving his Fourteen Points speech, Wilson had received an Inquiry report on 2 January 1918 which bore the title "A Suggested Statement of Peace Terms." Accompanied by a collection of about twenty maps, this report became the basis of Wilson's peace proposal (Martin 1980: 82–3).

An Inquiry memorandum of 13 November 1917, which lists the cartographic activity prior to the 2 January report, gives an indication of the types of maps which were submitted to Wilson.[45] No explicit mention is made of ethnographic or nationality maps, but a reference to maps portraying racial composition reveals an interest in cartographic information related to claims for national self-determination. As the personal narrative of the peace conference by US Secretary of State Robert Lansing shows (Lansing 1921: 96–7), race was commonly used at the time to connote nationality. Maps of the distribution of ethnic or national groups employed the same imprecise nomenclature. Often these maps were labeled "maps of peoples," "maps of nationalities," "maps of languages," and even "maps of races," when in reality the only data that were uniformly employed were surveys or official statistics of language use. The reason for the reliance on language data is simple: there were no other accurate data available for the production of such maps. Therefore, it is very likely that the second series of maps prepared by the Inquiry – which portrayed racial

15

compositions – was based on language data or that the maps were very general and small-scale.

There is also another explanation for the failure to explicitly mention ethnic or language maps: it was mainly Wilson who relied on the principle of national self-determination; the Inquiry was more apt to follow strategic and economic considerations (Gelfand 1976: 150–1). In fact, it might have been that Wilson relied so heavily on self-determination precisely because he did not have an accurate map of the distribution of nationalities in Central and Eastern Europe. If the president had seen the intermixed national composition of much of Eastern and Central Europe he might have realized that "just" territorial settlements along the lines of nationality and allegiance, which he proposed, were bound to lead to conflicts. Wilson's ignorance of the distribution of nationalities in Europe is revealed in the remark he made during the journey to Paris which was recorded by Charles Seymour, the Chief of the Austro-Hungarian Division of the American Commission to Negotiate Peace. Wilson expressed surprise that there was a great mass of Germans in northern Bohemia. "Why," he [Wilson] said, "Masaryk[46] never told me that" (cited after Seymour 1960: 108).

Later map work by the Inquiry included mainly the production of base maps for use at the peace conference in Paris. These maps and graphs could be used to document research results and to present proposals for territorial changes (Gelfand 1976: 91). In October 1918, when the Inquiry was at its maximum size, 17 of its 126 members worked in the cartographic division (Gelfand 1976: 46). The maps of the Inquiry were in great demand at the conference: nearly all delegations at Paris requested a series of base maps and block diagrams (Martin 1980: 83). Isaiah Bowman, the director of the American Geographical Society and Chief Territorial Specialist of the Inquiry, noted in his diary that maps were considered so important at Paris that it was said "one map is worth ten thousand words" (Martin 1980: 91).

Bowman's statement regarding the importance of maps at the Paris peace conference is markedly different from that of the director of the Inquiry, Sidney Edward Mezes:

> Base maps were constructed for the whole of Europe. . . . In volume this was one of the largest undertakings of The Inquiry. . . . But at the Conference these maps were hardly used at all. Some of the cases containing them were not opened. The world series of millionth maps proved to be sufficient for all needs.
>
> (Mezes 1921: 5)

Mezes's condemnation of the usefulness of the cartographic work of the Inquiry has to be seen in the light of the strained relationship between Mezes and Bowman. Bowman was in charge of the Inquiry's map work and during one period in 1918 seems to have usurped some of Mezes's powers. The rivalry went so far that Mezes even attempted to exclude Bowman from attending the conference (Gelfand 1976: 94–7, 354–7). The account of another peace conference participant, Charles Seymour, reveals that Bowman was probably more accurate in his assessment of the importance of maps:

> One of the most picturesque scenes of the Conference took place in Mr. Wilson's drawing room in Paris, with the President on all fours in front of a large map on the parquet floor, other plenipotentiaries in like posture, with Orlando crawling like a bear to get a better view, as Wilson delivered a succinct and accurate lecture on the economics and physiography of the Klagenfurt Basin. Maps were everywhere. They were not all good. Westermann refers to certain maps introduced by claimants in the Near East which it would be "a bitter derision to publish." But the appeal to the map in every discussion was constant.
>
> (Seymour 1960: 108)

The real worth of maps in the context of Wilson's Fourteen Points and the Paris peace conference went beyond the mere savings in voluminous documentation: maps became powerful political tools for territorial claims. Obviously, maps had been used to stake out what national groups considered the "true" extent of their national territory before the principle of national self-determination became officially recognized at the end of the First World War; but they were isolated incidences. Already in 1909, the Polish nationalist Roman Dmowski appended an ethnographic map to his book *La Question polonaise* to support his claims for an independent Polish state (Dmowski 1909).[47] And in September 1914, Serbian nationalists prepared a map of the future political organization of Europe on the basis of the national aspirations of the peoples of Austria-Hungary.[48]

A marked increase in maps of national territorial claims took place after the public affirmation of the nationality principle in the declaration of the war aims of the Entente on 10 January 1917. Ethnographic maps became regular features in the periodical *New Europe*, which was edited by R. W. Seton-Watson. The weekly acted as a forum for the political aspirations of the numerous national groups in Eastern Europe. The Czech nationalist and later president of Czechoslovakia Thomas Masaryk considered *New Europe* to be of the greatest assistance to the Czech cause (Masaryk 1927: 82). Some of the articles from *New Europe* were reprinted in other journals, such as the United States journal *The Bohemian Review*,[49] but the maps were not taken over wholesale. For example, the American reprint of Thomas Masaryk's *New Europe* article entitled "The Future Status of Bohemia" was accompanied by a slightly altered ethnographic map.[50] The Czech and Slovak territory in the American map edition extended in some areas beyond the one in the British edition. It claimed Czech and Slovak majorities east of the Troppau (Opava), which the Poles considered their territory. Obviously, the British weekly *New Europe*, although partisan

17

to the national liberation movements within Austria-Hungary, was not only more scholarly, but had to be restrained in its advocacy of specific territorial demands. It had to make sure that the nationalist claims of individual contributors did not arouse conflicts, e.g. between Poles and Czechs, and thereby lead to disunity.[51]

Later in 1917, the Czech nationalist Hanus Kuffner published a pamphlet with maps which argued that only a great Czech state in Central Europe would have the necessary military strength to ensure future peace. According to a 1922 translation of Kuffner's work (Kuffner 1922), the Czech original was distributed to the "pertinent groups" in the summer of 1917.

In addition to these explicitly partisan cartographic productions, the nationalist independence struggles in the First World War and the affirmation of the principle of national self-determination also stimulated a series of academic studies of the ethnic make-up of Europe. Examples are the work of the American Geographical Society staff member Leon Dominian (1917) and a three-part ethnographic map of Central Europe by the *Istituto Geografico de Agostini* (1917).

A veritable flood of maps outlining national territories on the basis of ethnic distribution came in the wake of Wilson's Fourteen Points speech. There are two reasons for this. First, Wilson explicitly linked national self-determination to justice. He explained in his speech that "an evident principle runs throughout the whole program . . . it is the principle of justice to all peoples and nationalities . . . whether they be strong or weak" (cited after Mantoux 1952: 36). Thereby, Wilson had made national self-determination a valid political program and a morally just basis for territorial claims.

Second, Wilson's speech promised impartiality and objectivity for the upcoming peace conference. He made it quite clear that territorial claims had to stand up to scrutiny – he spoke of "territories inhabited by indisputably Polish populations" (Gelfand 1976: 148) – which implied fairness and objectivity. This stress on objectivity meant a radical departure from the existing political practice in which career diplomats decided territorial settlements according to political expediency. Instead of old-style diplomacy, Wilson wanted to involve experts in history, economics, international law, and other subjects to establish "a new world order" (Shotwell 1937: 14, 28). The American State Department realized this "revolutionary, turning point in the history of American diplomacy" and worried that the Inquiry might actually pose a threat to the role of the State Department (Shotwell 1937: 12)

These two factors, Wilson's espousal of national self-determination as a just cause and his emphasis on scientific objectivity, had repercussions for maps. Maps have an aura of being scientific documents, of being accurate and objective (Hall 1981: 315), and they are presented as such by the authorities (Harley 1988a: 280). Their integrity is even greater if they are produced by respected scientists. In addition, claims of a national group for a certain territory can only be made comprehensible and effective if presented in graphic form. As a result, ethnographic maps became the ideal documents: they not only vividly showed the territorial claims but also substantiated them at the same time. Simple ethnographic maps turned into stakes for territorial claims in the upcoming peace treaties.

MAPS AND CLAIMS TO NATIONAL TERRITORY

National groups tried to make their territorial claims known in several ways: they addressed the Allied and Associated Powers directly; they published their claims in respectable journals; or they waited for the peace conference to submit their claims. Obviously, the most successful approach was to try to influence the commissions which were preparing the scientific basis for the peace treaties, in particular the American Inquiry.

The United States was not the only country which had set up a preparatory commission of experts. The governments of Great Britain and France established similar institutions which studied territorial and economic issues of the peace settlement (Shotwell 1937: 14). Yet the American institution had a more important role before and during the peace conference. The United States was the only major power which had no territorial interests in Europe and therefore could be considered impartial. In addition, the European states relied mainly on career diplomats in the preparation of the peace (Seymour 1928 III: 170–1) and contributed "men trained in their foreign offices and in their diplomatic corps" to the territorial commissions at the conference in Paris (Day 1921: 28). The Inquiry and its successor, the American Commission at Paris, were unique because they consisted of independent academics. Therefore, it was in the best interest of the national groups to influence the Inquiry or its successor, since a statement by these "neutral" American institutions carried more weight. At the conference in Paris, the findings of the American Commission were at times even accepted without question. Charles Seymour, who served as a US delegate on the Czechoslovak, Romanian, and Yugoslav territorial commissions, quoted the following statement by a foreign delegate: "I suggest that we accept the amendment without asking for the evidence. Hitherto the facts presented by the Americans have been irrefutable; it would be a waste of time to consider them" (Seymour 1921: 96).

However, even though the American Inquiry was professedly neutral and supposed to use scientific methods to ensure political independence (American Geographical Society 1919: 2–3), the activities of some of its members showed strong partisanship. Colonel House, who was behind the creation of the Inquiry, was close friends with the Polish nationalist representative Ignace Paderewski and repeatedly conferred with him on the issue of Poland (Gerson 1953: 67–85). At least since September 1918, Henry Arctowski and S. J. Zowski, working in the Polish Division of the Inquiry, kept Ignace Paderewski and the president of the Polish National Committee in Paris, Roman Dmowski, abreast of the activities of the Inquiry; the Polish nationalists adjusted their propaganda efforts in the United States accordingly (Gerson 1953: 96–7).

The existence of the Inquiry was revealed already on 29 September 1917, in a front-page story in the *New York Times* (America . . . 1917). In early 1918 even more information became public. A journalist (Arthur D. Smith) wrote a series of articles in the *New York Evening Post* (American Geographical Society 1919: 1–2) and also published a book which outlined the work of the Inquiry and its connection with the American Geographical Society (A. Smith 1918: 266–7). As a result, national groups with an interest in the future peace settlement arranged for conferences with the Inquiry staff and supplied information material (Wright 1952: 200). Maps were important in these endeavors to convince the Inquiry. James Shotwell, a member of the Inquiry and later the Chief of the History Division of the American Commission to Negotiate Peace in Paris, described one of a series of meetings between the Inquiry staff and representatives of national groups. At the meeting, which took place in a private dining room at the Columbia University Club, a presentation was given by the Czech nationalist and later president of Czechoslovakia Thomas Masaryk: "Professor Masaryk expounded the philosophy of self-determination for Eastern Europe, with the map which he had prepared hanging on the wall and the members of the Inquiry gathered like a class in a seminar" (Shotwell 1937: 10–11).

Czech propaganda in Great Britain during the First World War also used maps, as Masaryk pointed out: "Nor was our propaganda solely literary. We took a shop in Piccadilly Circus, one of the busiest corners of London, fitted it up like a bookseller's window, showed maps of our country and of Central Europe" (Masaryk 1927: 82).

Polish nationalists were similarly active and used maps to convince the Inquiry and the American government of their claims. Well informed by some members of the Inquiry, the Poles realized the importance of maps in the work of the Inquiry and the peace conference that was to follow.

The Inquiry collected a variety of information, such as census results, academic publications, and maps, and then used this material to prepare its own maps. While it was possible to submit additional statistical data to the Inquiry to convince its members of a certain boundary delimitation, it was clearly more effective to present a map by a renowned scientist which was based on the "correctly" interpreted data.

In 1918, A. Jechalski reissued parts of the 1916 *Atlas Polski* by the well-known Polish geographer E. Romer in the United States "on behalf of the friends and sympathizers of Poland" (Romer 1918). The maps depicted the historical territories of Poland, the distribution of Poles, and the landhold-ings of Poles in Lithuania and Ruthenia. On the frontispiece it carried a statement to the American public which presented the atlas as evidence of Poland's rightful territorial claims, and it also acknowledged that the United States and its President had pledged themselves to champion the cause of Poland. The US edition differed slightly from the Polish original (Romer 1916); it included relief which illustrated the strategic nature of some boundaries along easily defensible mountain crests.[52] At least in one case – the copy kept at the Memorial Library of the University of Wisconsin-Madison – an ethnographic map by the Polish National Committee was appended to the atlas.

The map of the Polish National Committee showed the distribution of Polish people (by district in percentages) and the historical boundaries of the Kingdom of Poland in 1772. It also proposed a new boundary line enveloping all areas with at least 20 per cent Poles and a few pockets with 5 to 20 per cent. The areas inside the proposed Polish territory were all shaded in reddish hues which were considerably darker than those beyond it. This made the argument convincing. The claim also appeared reasonable because a large part of the territory of 1772 was excluded. The correspon-dence between this proposed boundary line and the actual boundaries of Poland after the war is striking: except for the section of East Prussia which was given to Germany following the plebiscite, they are nearly the same.

During the peace conference in Paris, the Polish delegation published a booklet in March 1919 which contained essays by several Polish professors (Commission Polonaise 1919). On pages 29 and 31, respectively, of the booklet there was a map and accompanying description which argued that an intense effort of Germanization had taken place in parts of East and West Prussia, Poznan (Posen), and Silesia. The map allegedly showed the regional intensity of government-supported German colonization. As the text pointed out, government support was greatest in those districts with Polish predominance: "Where there are few Germans – therefore in regions which are essentially Polish – the percentage of Germans which are supported by the State increases" (Commission Polonaise 1919: 31).

Reports with accompanying ethnographic maps outlining territorial demands were common at the peace conference in Paris. In addition to the Poles, the Czechoslovak delegation submitted several memoranda,[53] and Serbian and Italian groups published studies to show their ethnic right to the contested area of Istria.[54]

Bulgarian nationalists published an ethnographic and historical atlas of the Balkan peninsula in 1917 to support their claims to all of Macedonia and part of Serbia (Rizoff 1917). The atlas was edited by the Royal Bulgarian envoy in Berlin, D. Rizoff, and contained forty maps with descriptions in German, English, French, and Bulgarian, which had been compiled by two professors at the University of Sofia.[55] Serbian academics were outraged, and A. Belic, a professor at the University of Belgrade, published an article in a periodical which was reprinted shortly afterwards in a series by the League of Serbian Universities (Belic 1918).[56] Belic condemned the atlas as a piece of Bulgarian propaganda. He analyzed and rejected the claims expounded in the atlas and pointed out the rightfulness of Serbian claims.[57]

The Germans, for obvious reasons, were not so well informed about the preparatory commissions toward the end of the war and would have had difficulty in presenting their case to the commission offices since they were located in enemy territory. Therefore, the German government focused attention on collecting material to be presented at the peace conference. But to their great surprise, they were excluded from the conference and only allowed to present counterproposals to the peace terms the Allies had agreed upon.

The German counterproposals were developed by the German delegation in Paris with the help of a group of experts in the Foreign Office in Berlin. However, neither the delegation nor the group of experts included geographers, that is, specialists on maps (Mehmel 1990: 55–6). Apart from government officials, the people involved in German peace preparations were mainly representatives of banks, industry, shipping, and private businessmen (Mehmel 1990: 56).

This is doubly surprising insofar as German geographers even volunteered their help. Erich Wunderlich, the director of the *Landeskundliche Kommission beim Generalgouvernement Warschau* from 1915 to 1919 (Roessler 1990: 275) and editor of the *Geographischer Bilderatlas von Polen*, offered his cartographic assistance to the German government already in March 1919 (Mehmel 1990: 58). On 10 May 1919, three days after the Allied peace conditions had been presented to Germany, the well-known German geographer Albrecht Penck submitted a series of newly prepared ethnographic maps to the government (Mehmel 1990: 58). The office in charge of German peace preparations had also received a publication with maps in three languages (German, English, French) by the geographer Ernst Tiessen (Mehmel 1990: 61). Yet, none of these offers appears to have had an effect on German peace preparations. Not even the official German memoranda which were taken to Paris were authored by geographers (Mehmel 1990: 63).

There are two possible reasons why the German government did not consider maps and geographic experts important. First, being ill informed about Allied peace preparations and being excluded from the conference, the Germans were probably unaware of how prevalent maps were and that many geographers participated on the Allied side. Second, the German government, in particular the Foreign Office in Berlin, did not expect any major changes in Germany's political boundaries and therefore ascribed only minor importance to territorial issues or maps (Mehmel 1990: 57; Grupp 1988: 292; see also below).

It is necessary to point out that even though German geographers had offered their help, they were ill prepared on questions of German national territory. Up until the summer of 1918, they were busy advocating German expansionism and justifying Germany's right to *Mitteleuropa*.[58] These territorial goals seemed within grasp after the treaties of Brest-Litovsk (3 March 1918) and Bucharest (7 May 1918). Only the collapse of the Western front in August 1918 and the German acceptance of the idea of national self-determination in the negotiations leading to the armistice of 11 November 1918 shifted their focus from a German-dominated *Mitteleuropa* to the delimitation of what should rightfully be considered German national territory.

This necessitated a departure from a concern with natural or economic regions and a new focus on German nationality or ethnicity; a topic which had been neglected up to that time by the scholarly community in Germany. The existing cartographic work, such as Paul Langhans's maps of the distribution of the German language (e.g. Langhans 1905), were not really useful. They were tainted by their association with the extreme nationalism of *völkisch* groups and they showed that Slavs were prominent in the eastern districts of Germany.

Wilhelm Volz, a geographer at the University of Leipzig, later admitted that German scientists had neglected the issue of national territory. In a letter of 2 July 1923 to the Foreign Office he stated:

> After the peace, German science strove to work for the German right in all endangered border regions of the empire by studying the ethnic distribution and by presenting its situation objectively. The present success could only be small because the fight started too late. . . . Who could have ever suspected earlier that at one time we would have to give proof of our right to extensive border areas − ? Thus, we entered the fight unprepared.[59]

GERMAN INITIATIVES

New maps were needed to argue for German territory and already in the winter of 1918, Albrecht Penck initiated a comprehensive mapping project which sought to depict the distribution of the German and Polish languages in the east of the German Empire. It involved students and at least seven academic staff members of the Department of Geography at the University of Berlin and the Institute for Oceanography in Berlin (Penck 1921a: 175–6). In a painstaking effort, they compiled the numbers of Polish, German, and bilingual speakers for 11,000 communities from published as well as unpublished results of the Prussian census of 1910. Under the "continuous direction" of Albrecht Penck, the cartographic assistant at the Institute for Oceanography, Herbert Heyde, represented the absolute figures in colored dots and squares on 116 sheets of the *Karte des Deutschen Reiches* (map of the German Empire) at the scale of 1:100,000 (Penck 1921a: 176).

Penck informed the public about this undertaking as early as 9 February 1919 in an article in the government-sponsored *Deutsche Allgemeine Zeitung* (Penck 1919a). Also in early 1919, a detailed description of the cartographic work appeared in a major German geographical journal (Häberle 1919). But the effectiveness of this project is questionable: it was not available in published form when the Allies discussed Germany's territorial provisions at the Paris conference. It was not even finished when the German government was presented with the conditions of the peace on 7 May 1919. Penck was only able to submit eighteen printed sheets, which covered the area of the lower Vistula and the Netze–Warthe region, to the German government on 10 May 1919 (Mehmel 1990: 58).[60] Even though Penck promised in his letter to the government that "the still remaining sheets will show the predominantly German character of Western Poznan" (cited after Mehmel 1990: 58), it appears that the government did not recognize the importance of the work and failed to give it financial support. A reviewer of the eighteen published sheets pointed out that high printing costs prevented the publication of the entire map and criticized the failure of government institutions, such as the census office, to produce such a work (Praesent 1919a: 220–1). He lamented:

> Was it necessary that the internal collapse of our fatherland had to occur in order for such an important and illustrative work for our domestic and foreign policy to be initiated and created? Could it not have been made long before to enlighten all those concerned *in time* and especially to be presented to our enemies so that they would not have needed to consult biased sources about the *true* distribution of nationalities? . . . Where is a similarly extensive and illustrative cartographic work for other disputed areas of our empire . . . ?
>
> (Praesent 1919a: 221; emphasis in the original)

Albrecht Penck also published a small-scale map to illustrate his article in the *Deutsche Allgemeine Zeitung* (Penck 1919a). The map was badly designed because it employed inconsistent graphic categories for the German and Polish majorities (see figure 2.1). The category *"rein deutsch"* (purely German) in the legend appears to belong to the Polish categories. Therefore, the map gives the impression that all areas identified by cross-hatching or a black hue are Polish. Despite these flaws, the local representatives (*Volksräte*) of the districts of Poznan (Posen) and West Prussia considered it to be the best representation of the German standpoint at the time and reproduced it in their weekly report of 14 March 1919 (Materialien . . .

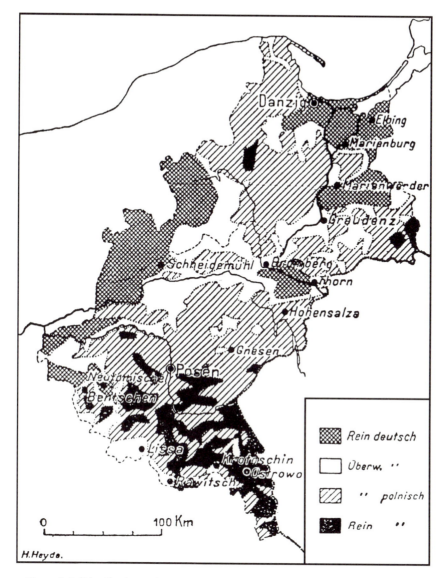

Figure 2.1 Distribution of Poles and Germans in West Prussia and Poznan I
(Penck 1919a).

1919: 2–3). The map was also published in a pamphlet by the *Reichsverband Ostschutz* (Praesent 1919a: 221).

The haphazard and ineffective character of the initial German initiatives is revealed in the way Penck's (1919a) map was modified by other German geographers. D. Häberle (1919) attempted to improve on

25

Penck's map and added a further category for the Kaschubes, a Slavic ethnic group that lived near the Baltic coast west of Danzig (Gdansk). He used dots to identify the Kaschubes, which when added to the hachures used for the Polish areas made them look very much like the category "purely German." Also, he continued to use Penck's other flawed categories. R. Reinhard (1920) altered the design of Penck's categories: German majorities were identified by black hue and cross-hatching, Polish majorities by light shading and dots. He thereby avoided the shortcomings of Penck's map, but he failed to add the Kaschubes, which would have made the German argument stronger. This is doubly surprising, because he mentioned in the text that the Kaschubes were different from the Poles and that it was incorrect to cite them in support of Polish claims (Reinhard 1920: 11–12).

Another German academic, Dietrich Schäfer, became active right after the armistice and prepared a map of the distribution of the German language at a scale of 1:1,000,000.[61] In a review in a major German geographic journal this less ambitious cartographic work was labeled as "crude" but also as "a skillful counterpart to similar [maps] published by the Polish side" (Praesent 1919b: 128). Thus identified as being tendentious by German geographers, it is unlikely that it had any effect abroad.

Germans were not only late in addressing the issue of what constituted German territory, but much of their activity was also reactive, that is, German academics rarely took the initiative without feeling challenged by foreign cartographic publications. For example, Stahlberg (1921: 10) states that Schäfer presented a map in November 1918 to contradict the claims to German territory made by a "falsified" Polish map.[62] A German historian, Erich Keyser, attacked a map published in Switzerland in French by Josef Freilich.[63] To discredit Freilich's map, he published a small booklet and a color map in 1919, which argued the rightfulness of German claims to the province of West Prussia (Keyser 1919). The booklet was not part of an academic series or other periodical, and it is not clear for which audience it was intended. It is possible that Keyser submitted it to official German institutions for use as evidence in peace negotiations.

The majority of German cartographic works dealing with German national territory in the late 1910s and early 1920s was sparked by a map which first appeared as a German product, but which was later revealed as partisan to the Polish cause. It was Jacob Spett's map of nationalities in the eastern provinces of the German Empire.[64] The map showed that a clear Polish majority existed in the area of West Prussia and Poznan, thereby substantiating the demands of the Polish National Committee for Polish access to the Baltic Sea. It appeared to be a genuine German production: the text was in German and on the front of the map, at the bottom, it said: "Gotha: Justus Perthes." Without comment, Häberle (1919: 124) mentioned it together with other ethnographic maps from Germany and Austria, and Praesent even considered it "a desirable condensation and replacement of the similar well-known individual maps of the eastern provinces by P. Langhans" (Praesent 1919b: 129). But on 19 March 1919, Spett's map quickly fell from grace: the French newspaper *Le Temps* published a section of it to show that the Polish demands were justified since they were even supported by the products of German science.

German geographers were outraged and responded immediately. They insisted that the map was not a German product and called it a "masterpiece of deception" (Penck 1919b: 537). The map had only been printed by Justus Perthes but was really published in Vienna by Moritz Perles. Penck also responded cartographically to Spett's map and published his view of what constituted Germany's rightful territories in the East in the Leipzig illustrated periodical *Illustrierte Zeitung* (Penck 1919b) and even a color version in a leading German geographic journal, the *Zeitschrift der Gesellschaft für Erdkunde zu Berlin* (Penck 1919c).

Penck's maps used a novel form of representation, a variation of the dot map, and depicted the ethnic distribution in West Prussia and Poznan in absolute figures. The design of these maps was much better than his earlier attempt (Penck 1919a). They showed that there were no Poles in the coastal area and that the remaining areas were thoroughly intermixed with a clear German majority along the Netze river. This was achieved by choosing symbols for the Kaschubian majority which were very similar to those for the Germans. For example, the black and white version in the *Illustrierte Zeitung* used black dots to designate Germans, white circles with a black rim for Poles, and black-rimmed dots with a black center for the Kaschubes. As a result, the symbols for the Kaschubes and Germans seemed to belong together and were associated by the map reader (see figure 2.2). The color version in the *Zeitschrift* was even more misleading. Blue dots designated the Germans and red dots the Poles, while the Kaschubes were represented by blue-gray dots which were virtually indistinguishable from the German symbols.[65]

The late start of German cartographic activity with regard to their territorial claims and the customary delays in publishing rendered the German influence on the Allied and Associated Powers ineffective; they simply did not reach them before the peace conditions were worked out.[66] Had they dealt with these issues earlier, the German geographers would have been heard. German geography had a worldwide reputation for impeccable scholarship, and German geographic journals were consulted by academics in many European countries and in the United States. Since the United States, France, and other Allied countries had geographers on their peace commissions, German geographic publications would have been noticed.

It is ironic that with all the attention German academics devoted to refuting foreign maps, they apparently failed to notice the pro-Polish stance

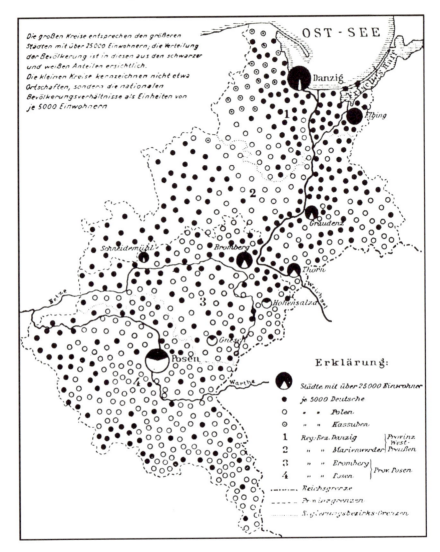

Verteilung der Deutſchen und Polen in Weſtpreußen und
Poſen. Nach einer Darſtellung von Prof.Dr.Albrecht Penck.

Figure 2.2 Distribution of Poles and Germans in West Prussia and Poznan II
(Penck 1919b).

of ethnographic maps in most German atlases until other countries pointed
them out. These small-scale German maps showed a distribution of ethnic
groups similar to Spett's and Freilich's maps and thus supported the
creation of a Polish corridor to the Baltic Sea. There is evidence not

only that the Inquiry consulted these German atlases,[67] but that the French were aware of them too. The French prime minister Briand referred to the German atlas *Andrees Handatlas* of 1914 as evidence for the Polish claims to Upper Silesia (Stahlberg 1921: 6). Rather meekly, the German academics pointed out that the ethnographic maps in German atlases were too simplistic to be used for the delimitation of national boundaries. They did not designate minorities, such as the Kaschubes and Masures, and only showed areas with absolute majorities, not the intermixed distribution of Germans and Poles (Penck 1919b: 536; Stahlberg 1921: 7). Stahlberg believed that these simplistic representations existed in German atlases because Germans were "apolitical" and therefore failed to see the political significance of ethnographic issues (Stahlberg 1921: 7).

Faced with the break-up of the multi-ethnic Dual Monarchy, geographers from Austria-Hungary also embarked on projects after the signing of the armistice in November 1918. Austrian-German geographers who had been very active in propagating *Mitteleuropa* ideas shifted their attention to self-determination, in particular the *"Anschluß"* with Germany (Meyer 1946: 189). Only a month after the armistice, on 10 December 1918, Rudolf Laun, a professor at the University of Vienna, had finished a pamphlet with two maps which outlined ethnic German territories in Bohemia and Moravia (Laun 1919). The pamphlet was intended as a contradiction of Czechoslovak demands.

In early 1919, under the direction of the geographer Robert Sieger, the University of Graz issued the memorandum *Die Südgrenze der deutschen Steiermark*. It argued that the ethnic mix in Styria made it necessary to use natural landscape boundaries (Hassinger 1919: 218). The Hungarian Geographical Association initiated a comprehensive large-scale mapping effort and prepared an ethnographic atlas using a method similar to the project initiated by Penck in winter 1918 (Penck 1921a: 175).

There appears to have been little cooperation among the academics of the Central Powers during this period. Germans and Austrians generally agreed on the desirability of the *"Anschluß,"* but as the Austrian geographer Hugo Hassinger (1919) pointed out, many Germans were ill informed about the "true" boundaries of the ethnic German territory in Austria-Hungary. He criticized the exclusion of Austrian-German territories from an ethnic map in the *Illustrierte Zeitung* (Leipzig) and also from another map which accompanied a memorandum which the German geographer Walter Vogel had submitted to the German government (Hassinger 1919: 216).[68] Hassinger feared that such maps could lead German politicians voluntarily to cede territory which rightfully belonged to Germany (Hassinger 1919: 217).

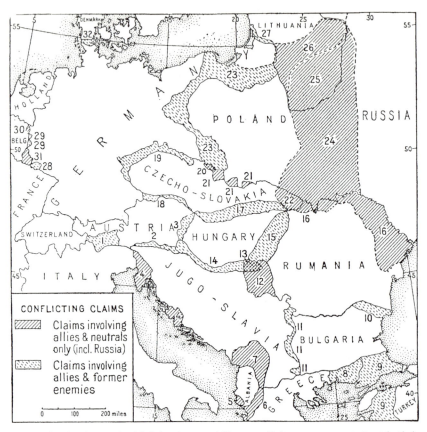

FIG.   Overlapping territorial claims in central Europe.   Claims are represented not in their most extreme but in their more conservative forms; in general, therefore, the ethnic line is taken as the limit of the claims of Austria and Hungary; the eastern limit of Poland's claim as shown on the map is some distance west of her boundary in 1772.   The districts are numbered as follows:

1. Part of Austrian Tyrol
2. German-Slovene borderland
3. German Hungary
4. Istria and Dalmatia
5. Valona
6. Northern Epirus
7. Serbo-Albanian zone
8. Western Thrace
9. Eastern Thrace and the area claimed by Greece in Asia Minor
10. Southern Dobrudja
11. Western Bulgaria
12. Southern Banat
13. Northern Banat
14. Southern Hungary
15. Western Transylvania
16. Eastern Ruthenia and Bessarabia
17. Southern Slovakia

18. Southern Bohemia
19. German Bohemia
20. Czech districts in German Silesia
21. Teschen, Orawa, and Spits (named in order from west to east)
22. Ruthenia
23. Upper Silesia, Posen, Danzig, Marienwerder, and Allenstein
24. Polish-Russian border zone
25. Lithuanian-Polish-Russian border zone
26. Polish-Lithuanian border zone
27. Trans-Niemen territory
28. Saar basin
29. Malmédy, Eupen, and Moresnet
30. Southern Limburg
31. Luxemburg
32. Northern Slesvig

Figure 2.3 Conflicting territorial claims in Central Europe (Bowman 1922: 4).

THE PEACE TREATIES OF PARIS

The peace conference in Paris faced the difficult task of balancing the conflicting territorial demands of the numerous national groups of Europe (see figure 2.3) and the interests of the Entente Powers as outlined in secret treaties. Although Wilson's Fourteen Points and the armistice negotiations had emphasized the principle of national self-determination, its application to the drawing of boundaries proved to be extremely challenging. Not only was the distribution of nationalities in Central Europe intermixed, but there were also strategic and economic aspects which had to be considered to ensure that the newly created states were viable. As a result, the territorial provisions of the Paris peace treaties deviated from the nationality principle, and several states included large minority populations.

For example, to ensure easily defensible boundaries, Czechoslovakia encompassed a large German population in the mountains of western Bohemia, a territory later loosely called Sudetenland. As promised in the secret Treaty of London, Italy was given her "natural" or "geographical" northern frontier in the Brenner Pass region, which effectively placed an area with 230,000 German speakers under her jurisdiction (Bowman 1922: 134). The spatial nomenclature for this area reveals the propaganda power of semantics. Austrians called it South Tyrol, which evoked an inextricable link with Austrian North Tyrol, while the Italians named it Alto Adige to show that the area was complementary to the lower Adige valley and therefore an intrinsic part of Italy (Hall 1981: 321).

In the east of Germany, the districts of Poznan (Posen) and West Prussia which had mixed Polish–German populations were included in the recreated Polish state, and the German city of Danzig was internationalized. This provided Poland with access to the Baltic Sea as demanded in Wilson's Fourteen Points. Citing the right of national self-determination, German Austria had expressed the wish to unite with Germany as early as 12 November 1918, but the "*Anschluß*" was forbidden in the Paris peace treaties (Grupp 1988: 300).

The Treaty of Versailles, which took effect on 10 January 1920, outlined territorial losses, restrictions on military strength, and payment of reparations.[69] The final boundaries of Germany which resulted from the treaty are illustrated in figure 2.4.

In Germany, the Versailles Treaty was received with outrage. The German public and all political parties from the left to the far right were convinced that it was unfair and unjust and claimed that it violated the right of national self-determination. Politicians, the press, and the general public were united in their whole-hearted rejection of the treaty and expressed their anger in countless protest marches, rallies, and written exposés.

Yet, most of the provisions of the treaty which the Germans opposed so vehemently, such as the creation of a Polish corridor through Germany,

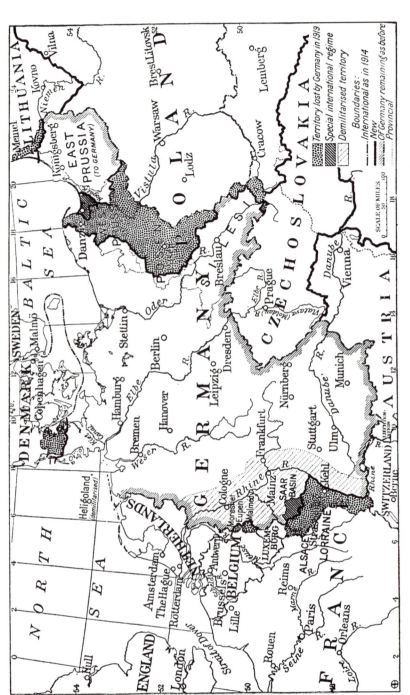

Figure 2.4 Territorial provisions of the Treaty of Versailles (Bowman 1928: 274).

were in agreement with Wilson's Fourteen Points and therefore should have been expected. What had happened was that the Germans had closed their eyes to any unfavorable interpretation of Wilson's Points. The general public in Germany was simply misinformed: all through the war the military dictatorship of Ludendorff and Hindenburg had led them to believe that victory was certain. Especially after the vast territorial gains of the Treaty of Brest-Litovsk on 3 March 1918 the promises of impending victory seemed beyond doubt. Even members of the civilian government, who should have known better, held steadfast to this illusion of a victorious or at least favorable conclusion of the war. Only for a brief period before the acceptance of the Fourteen Points as a basis of the armistice did the German government have doubts about the benefits of the Fourteen Points for Germany. The chancellor of Germany, Max von Baden, spoke out against a submission to the Fourteen Points and only wanted to accept them if they were reinterpreted by Germany (Grupp 1988: 292).

Soon afterwards the correct assessment that Wilson's Fourteen Points would not be entirely favorable to Germany was abandoned by the German government. Already on 6 November 1918, a member of the German Foreign Office noted that it was now certain that a peace on the basis of the Fourteen Points would not result in a weakened Germany (Grupp 1988: 292). The German government never abandoned this illusory belief and only prepared for the unrealistic case of an entirely pro-German interpretation of the Fourteen Points; they failed to deal critically with the practical application of Wilson's theoretical program (Mehmel 1990: 56–7). Reality hit when the peace treaty was presented. Measured against the spurious hopes of the Germans, the provisions appeared harsh indeed.

The German belief in the injustice of the Versailles Treaty fueled attempts at its revision. These were mainly directed at the new eastern boundary, which contained the most disputed provisions (Jessop 1942: 70; Batowski 1970: 198). Maps figured prominently in revisionism: maps were the only effective means to establish a clear German consensus on national territorial claims and maps were a perfect medium to present a persuasive argument to the Allied Powers. As a consequence, the discourse on German national territory became cartographic.

3

A CONCERN FOR ACCURACY

Thus, the untruthfully constructed map became a prostitute, raped in the national and literary sense.

(Eckert 1925: 146)

After the conditions of the peace treaty became known, German scientists devoted their energy to ethnographic mapping in earnest and analyzed maps that supported the peace conditions, namely Spett's 1918 map and Romer's 1916 atlas. On the basis of these analyses, they developed more accurate representations, which outlined the ethnographically "just" boundaries of Germany. The discrepancy between the existing boundaries and those of the improved maps were supposed to document the injustice that had been done to the Germans. Scientific research turned ethnographic maps into documents for revisionist demands.

The initiators of the use of scientific cartography towards this end were almost all geographers. True to the scientific philosophy of the time, positivism, they nevertheless believed that their activities were not biased. Since they worked scientifically, they deemed the results of their work to be absolutely objective (Geisler 1926: 712).

EXAMINATION OF EXISTING MAPS

Entirely convinced that the territories Germany had to cede in the Treaty of Versailles were "truly" German, German academics sought explanations of how this could have happened. The answer was clear to them as soon as the Parisian journal *Le Temps* published a section of the Spett (1918) map on 19 March 1919 (see figure 3.1). As Albrecht Penck wrote in May 1919: "This masterpiece of forgery . . . probably contributed to the unacceptable conditions of the peace treaty, on the basis of which more than two and a half million Germans are to be separated from the empire" (Penck 1919b: 537).

Many German authors shared this view (Heyde 1919, Stahlberg 1920 and 1921, Geisler 1926 and 1933); they were convinced that falsified maps had

Figure 3.1 Section of Spett's 1918 map (*Le Temps*, 19 March 1919: 2).

been presented to the Allies at Paris. The chief culprits in their eyes were the Poles. They exposed the "falsifications" or design flaws in maps supporting the Polish case and then offered alternative solutions.

Despite nationalist biases in their interpretations, German geographers and cartographers had a keen understanding of the general limitations of maps. They pointed out that variations in the map content of several maps could be attributed to differences in the data or to the method of representation.[70] They stressed that the methods of representation were directly related to the scale and that they should not be interchanged between general or overview maps and detailed maps.

On the basis of these three premises, data, method of representation, and scale, they analyzed the non-German maps that supported the post-Versailles boundaries. Various maps received attention, but the two main examples which were cited were the maps in Eugenius Romer's Polish atlas (1916) and Jakob Spett's (1918) map.[71]

German geographers argued that non-German maps were based on the wrong data. The maps had not been drawn relying on existing census results such as the Prussian census of 1910. Instead, the official data had been falsified or augmented with other questionable ethnographic surveys, such as the *Schulstatistik* of 1911.[72] This latter survey was deemed biased by the German authors because the data had not been collected by independent officials, but by schoolteachers. The German argument went as follows: the teachers might have increased the number of Polish-speaking children in their classes in order to make it appear that their instruction posed great challenges, which in turn would enable them to ask for more money (Penck 1921a: 183). The survey only included teachers from the lowest level of schools, where Germans were clearly under-represented because of their "greater affluence and their higher social position" (Geisler 1933: 47). In addition, the Poles had more children than Germans, which also introduced a bias (Penck 1921a: 184).

For German scientists there was only one reliable data source for ethnographic maps: the Prussian census of 1910. Although they admitted that the census had some flaws, they deemed them to be minor since they had been identified and discussed before;[73] it certainly did not warrant the label of falsification which the Poles had ascribed to the census (Penck 1921a: 183). There was a good reason for the German reliance on the 1910 census: no other statistical data offered more favorable results for Germany if interpreted "correctly."[74] A key feature of the 1910 census was that it listed not only Germans and Poles, but also the Slavic people called Kaschubes. This was of central importance to Germans since the Baltic coast region of the corridor that had been given to Poland was inhabited by a Kaschubian majority. It led to the German claim that the Polish access to the Baltic Sea could not be justified on the basis of Polish self-determination. As Penck pointed out: "A Polish corridor to the [Baltic] Sea in the sense of a language area has never existed. It is a slogan for political aspirations, which ignores the difference between Poles and Kaschubes" (Penck 1921a: 184).

There seemed no doubt in the mind of German scientists that the

Kaschubes were linguistically different from the Poles. Penck mentioned that in census surveys most Kaschubes had listed Kaschubian as their mother tongue, rather than Polish, and he even cited Polish scholars in support of his argument (Penck 1921a: 181–2). Yet, only two years earlier Penck himself had stated that the Kaschubes were closely related to the Poles (Penck 1919a).

Both Romer's atlas (1916) and Spett's map (1918) subsumed the Kaschubes and other linguistic groups, such as the Masures – which the Germans also considered to be an independent people – under the Poles. While Romer was criticized for augmenting the Prussian 1910 census with the "biased" school survey of 1911 and his own research data, Spett was believed to have vindictively falsified his data. In fact, Spett's map was labeled a "scandalous falsification" (Geisler 1926: 707; Heyde 1919: 185). Although Spett's map explicitly mentioned that it was based on the Prussian census of 1910, Herbert Heyde (1919: 185) pointed out that every one of the sixty-six random checks he ran on Spett's map deviated from the official census figures.

Geisler (1926: 705) insinuated that Romer, who was a renowned Polish geographer, had also intentionally altered the data for one of his maps, and questioned why Romer had used Spett's map as a source for the province of Poznan but not for West Prussia. According to Geisler, this selective use of different data served to "obfuscate his arbitrary deviations from official figures and to fake scientific accuracy" (Geisler 1926: 705). The problem with Geisler's argument is that he accused Romer of using a map which was published two years *after* his atlas. Geisler even stated that Romer had listed Spett's map as a source in his atlas. As is to be expected, Romer's 1916 atlas does not contain any such assertion, which only leaves the possibility that Geisler analyzed one of the later editions, none of which appeared before 1919. However, since Geisler claimed to reveal falsifications of maps to help explain why Germany had lost national territory at Versailles, his use of a post-Versailles edition of Romer's atlas undermines the credibility of his discussion of Romer's work.[75]

A further point of criticism was the bilingual population. Penck (1921a) believed that they ought either to be treated as a separate category or to be distributed evenly between Poles and Germans.[76] He found it unacceptable that Polish maps presumed bilingual speakers to be ethnic Poles who knew German but had been forced to admit bilingualism (Penck 1921a: 182).

German scientists also deemed the methods of representation of Polish maps to be misleading. They admitted that the color schemes of Polish maps supported the Polish cause very effectively, but that they did so at the expense of truthfulness. For example, Romer's map of the distribution of Poles in his 1916 atlas used two shades of blue to designate the categories less than 1 per cent Poles and 1–5 per cent Poles, but for the remaining categories of 5–25 per cent, 25–50 per cent, 50–75 per cent, and above 75 per cent, the map used shades of increasing intensities of brown.[77] This gave the impression that all areas with more than 5 per cent Poles were a unified group and that Polish majorities extended along a corridor to the Baltic Sea (Geisler 1926: 705).

Spett's map was also considered to be sophisticated in its misleading use of color. It appeared to favor the Germans because it displayed them in red, a very dominant color. But according to Geisler (1933: 54) a closer inspection revealed that Spett chose a red that was almost violet. This made the Germans look less prominent than the Poles, who were designated by a bright green (Geisler 1933: 54).

The effectiveness of color schemes in Polish maps was contrasted with the shortcomings of German maps. Paul Langhans's 1905 map was rejected because its colors were so subdued that only areas with more than 75 per cent Germans were clearly visible.[78] In addition, the areas with less than 25 per cent Germans were not given a color but simply left white, which made it appear as if no Germans lived there (Geisler 1926: 707; Geisler 1933: 54–5).

Although Langhans's 1905 map was thought to have an inadequate color scheme and failed to depict the Kaschubes – a major flaw according to German scientists – its data representation was considered to be truthful and reliable. In the eyes of the Germans this was not the case with Polish maps. They noted that most Polish ethnographic maps had unsystematic and biased class intervals. Romer's 1916 atlas map of the distribution of Poles used intervals of 25 per cent which would have been acceptable had he not also included two further subintervals within the first 25 per cent. According to Geisler (1926: 705) these additional intervals, which designated below 1 per cent Poles and 1–5 per cent Poles, revealed the "agitatorial-propagandistic" intention of the map. Spett's 1918 map was also criticized for having irregular intervals of 15, 30, 50, 70, and 85 per cent, but even more for its inconsistent representation of forest areas. Forest areas in his map were apparently only designated as uninhabited when they were inside a region with a German majority (Heyde 1919: 186; Geisler 1926: 709; 1933: 55–8).

A further point of contention for German scientists was that Polish maps employed methods of representation which were not deemed suitable for the scales of the maps. Penck (1921a: 174) rejected Romer's use of isarithms (= lines connecting points of equal value) to portray the distribution of Poles as percentages of the total population on plate IX of his 1916 atlas. He argued that isarithms were useful for the depiction of elevation levels but not for languages or peoples which had an intermixed distribution. However, he added that isarithms were a possibility for overview maps at a small scale. Since Romer's map was at a scale of 1:5,000,000, it is not quite clear what Penck intended to say.

German scientists devoted most of their attention to one method of representation, choropleth mapping. In this method, administrative units are colored or shaded according to the percentage quotas of language use. They felt that these representations were misleading on large-scale maps, since they did not take population densities into account, which could be extreme, especially between urban and rural districts. Geisler (1933: 52) compared two regions with a similar size, the area around Danzig and the Tucheler Heide. Both had clear majorities, the Danzig region was predominantly German and the Tucheler Heide predominantly Polish. He argued that it would be unfair if the 440,000 inhabitants of Danzig were to receive the same areal percentage value as the 55,000 inhabitants of the sparsely populated Tucheler Heide. German scientists also pointed out that areal coloring gave the impression that borders between different majorities were clear-cut along administrative boundaries, whereas in reality the distribution changed gradually (Penck 1921a: 172–5). Spett's large-scale map (1:500,000) not only used areal coloring, but according to Geisler (1933: 58–9) even altered administrative boundaries in some cases to reduce German majorities.[79]

However, German scientists thought that choropleth maps could still be used if they were modified. Penck (1921a: 171) found it acceptable to depict mixed areas by shading in the colors of the different languages, but he rejected the use of irregular minority-language "islands" inside the majority area. In his view, this gave the erroneous impression that well-defined geographical language regions of the minority existed instead of the actual intermixed distribution. As an example of such a misuse, he mentioned the map "Haack-Hertzfeld, *Wandkarte der Völker Europas*, Gotha: Justus Perthes."[80]

Although the first accusations of Polish map falsifications were made by German scientists as early as 1919, it was not until 1933 that the German government became involved in exposing this apparent injustice. In 1933, the Foreign Office entrusted Walter Geisler, then a professor of geography at the University of Breslau, with a "fundamental scholarly study" of the Spett map.[81] Geisler published a 76-page monograph with five large fold-out color maps to show the errors in Spett's map. The publication was part of the well-known geography series *Petermanns Mitteilungen, Ergänzungshefte*.[82]

The idea behind the publication was to address the scientific world through a scholarly periodical which was "above any suspicions of propagandistic tendencies."[83] But it was not left to chance that the monograph reached its proper audience: the Foreign Office ordered its representatives abroad to bring Geisler's work to the attention of select, well-known foreign scientists.[84] In March 1934, the monograph was already out of print, and a second edition was published. This proved to be fortuitous, since the original print contained errors as a result of improper color selections.[85] There was also a plan for a popular edition, but it seems that it was not carried out.

The question arises: why did the German government wait so long to take action? One possible reason was to avoid embarrassment. As Walter Stahlberg was eager to point out in his 1920 article, Spett's map had passed the official German censors (Stahlberg 1920: 770). Stahlberg specifically blamed the Foreign Office and ridiculed it by explaining that the customary abbreviation for the Auswärtiges Amt, "AA", stood for "*ahnungslose[s] Amt*" (ignorant office). Needless to say, Geisler extenuated the censorship incident and did not even mention the Foreign Office. Instead, he put the blame squarely on the Poles and on Spett, emphasizing that Spett was a Polish Jew who had moved from Vienna back to Poland after the end of the war. Geisler also excused the involvement of the publishing house of Justus Perthes in the printing of the Spett map and pointed out that there was no reason for Perthes to suspect anti-German intentions since Spett was a government official working for the state railway of Germany's wartime ally Austria (Geisler 1933: 14). His benevolent treatment of Perthes – which was in contrast to Stahlberg's (1920: 769–70) sharp criticism of the publisher – is easily explained. Geisler's 1933 work was not only supported by government institutions (*amtliche Stellen*), but also by Perthes itself.[86]

Another explanation for the paucity of official German involvement with regard to Spett's map is that the German government simply did not pay much attention to maps, just as had been the case during the peace conference preparations. It waited for an impulse from outside; it did not take the initiative. It was reactive rather than active.[87] The incentive was the use of the Spett map in Polish propaganda. Geisler (1933: 13; 1939: 554) stated that the motive for his monograph had been the publication of a reduced copy of Spett's map in the illustrated Polish periodical *Swiatowid* and the accusation by the Poles that Germany had confiscated this map. Correspondence by the Foreign Office also mentioned that Geisler's monograph was intended as a response to Polish propaganda.[88]

The reluctance of the German government to exploit the Spett map for propaganda is also revealed in the following episode. In the summer of 1935, the Spett map was shown at an oceanic exhibition in Cracow, Poland. Regierungspräsident Budding of Marienwerder (East Prussia) asked the German Ministry of the Interior (*Preußisches und Reichsministerium des Innern*) to suggest to the Foreign Office that it take action against the display of Spett's map. Despite repeated appeals by Budding that the Foreign Office file a protest with the Polish government, there are no records that action was taken.[89]

PRODUCTION OF NEW MAPS

German scientists did not just point out the falsifications or errors in maps that might have been used at the peace conference. They also presented improved versions of ethnographic maps. They claimed that these maps avoided the shortfalls of the existing maps and that they constituted the most accurate depiction possible. Albrecht Penck was again the precursor. He presented an ethnographic map of the eastern borders of Germany at a scale of 1:300,000 to the scientific world in a scholarly journal (Penck 1921a).[90] The map was based on the Prussian census of 1910 and constituted an outgrowth of his cartographic activities at the end of the First World War in preparation for the peace (see Chapter 2, "German Initiatives").

Penck's 1:300,000 map – which covered the entirety of the northeastern territories ceded in the Treaty of Versailles, that is, the so-called corridor – showed a complex distribution of language groups. Most of the map was covered by mixed language areas. Purely German areas were in the west and northeast, and a German "bridge" extended across the corridor along the Netze river and part of the Vistula river near Thorn. By comparison, purely Polish areas were much more limited and only encompassed groups of Polish villages, none of which had more than 2,000 inhabitants (Penck 1921a: 179). In addition, his map showed that a Polish corridor to the Baltic Sea did not exist since the coastal area had a Kaschubian majority (Penck 1921a: 180–4). Therefore, in Penck's eyes, the Treaty of Versailles fell short of realizing national self-determination and he predicted that the borders would not remain (Penck 1921a: 185).

Penck used the new method of representation he had developed for his 1919 map (Penck and Heyde 1919), which he called "absolute representation." This variation of a dot map attempted to depict the absolute number of individuals for each language group at their respective geographical locations. He identified each of the following language groups by color: Germans, Poles, Kaschubes/Masures, Bilinguals, and Speakers of Other Languages (*Anderssprachige*). The symbols of his maps were color-coded for the different language groups and ranked according to size: colored dots represented twenty members of each language group and colored squares of increasing size represented 200, 2,000, and 20,000 members.[91]

Penck pointed out that the application of his "absolute method" ensured a true representation of the precise geographical distribution of the different languages. He deemed it possible to achieve the same results for very small scales and referred to his colored map in the 1919 volume of the *Zeitschrift der Gesellschaft für Erdkunde zu Berlin* (see figure 2.2 for a black-and-white version of this map). The hue produced by the spatial agglomeration of the colored symbols made the language boundaries visible without the need actually to draw them (Penck 1921a: 176).

Other German scientists did not agree. Max Eckert (1925: 468) argued that Penck's method was inconsistent and arbitrary in the way it grouped the symbols to represent settlements of different sizes: settlements with up to 1,000 inhabitants were displayed with dots alone, but those which were slightly larger were depicted already with squares for 200 inhabitants. In addition, his maps were considered to be difficult to read and to give the impression that each community had its own separate existence (Eckert 1925: 468–9; Geisler 1926: 709). Another critic pointed out that Penck's maps portrayed the dispersion rather than the geographical grouping of the different languages, which made it very hard to base boundary delimitations on them (Milleker 1937: 640). Despite these shortcomings, Eckert (1925: 468) still believed that Penck's two large-scale maps were accurate graphic depictions which could be used to support German claims to the corridor.

This was Penck's intention: to produce evidence for German revisionist demands. To show the advantages and greater accuracy of his new method, he even had his assistants draw up comparative maps. They were based on the same data, the Prussian census of 1910, but they used the representational methods of the Polish maps, areal coloring and isarithms. In his view this provided extensive material for a more precise delimitation of the language boundaries and for a better evaluation of the usefulness of the different methods of representation. He stated that the maps were kept at the geographical institute of the University of Berlin and were available to the scientific community (Penck 1921a: 176 fn. 8).

Penck's new method soon attracted interest in Germany. In 1921, the *Preußisches Statistisches Landesamt* published a map of the plebiscite results in Upper Silesia at a scale of 1:100,000 which applied the same principles.[92] Leo Wittschell from the geography department at the University of Königsberg extended Penck's work on the corridor area and made a map of the distribution of nationalities in the area south of East Prussia.[93] It was an exact copy of Penck's method (Geisler 1933: 52), even though it employed plebiscite results instead of language data. Penck's method was even adopted for use in a popular publication (Pomoranus ca. 1931).

Wilhelm Volz, a geographer at the University of Breslau, also developed a new method, which, however, failed to find imitators. First applied in 1922 in a map of Upper Silesia,[94] it was only employed afterwards by Volz and his disciples. The method was described rather awkwardly as follows in the text accompanying a 1925 map of the distribution of Sorbs and Germans in the Upper and Lower Lausitz:

> The area of study is divided into large squares. In our case, squares with a side length of 5 cm (= 10 km), that is, with a coverage of 100 square km, were selected. The settlements and individual farms within one square are calculated according to the number and language of its inhabitants and the number of inhabitants in the large squares are

combined and represented in areal coloring, and to be more precise, in our case, in such a way that a square with a side length of 5 mm represented 150 inhabitants. Therefore, a population of 15,000 within 100 square km will result in a uniformly colored square. The distribution of the [different] languages is indicated by [different] colors.

(Zur Bevölkerungskarte . . . 1925: 160)[95]

The resulting map was a patchwork of differently colored squares which was superimposed on a topographic map. Maps using Volz's method looked very unusual and were not immediately intelligible without an accompanying explanation. Since the colored squares always extended beyond the actual settled area on the map, the placement of the different language groups was dependent on the availability of space on the map, and not on their most accurate location. This fact, together with the patchwork distribution, did not facilitate the drawing of language boundaries. In addition, towns with high population densities on the map had to be represented with a different symbol, such as a circle with sectors indicating the proportions of the different language groups, in order to avoid the colored square for the town covering the entire region and thereby completely obfuscating the language distribution in adjacent smaller settlements. These technical problems, the strange appearance, and the confusing description of his method might have dampened the enthusiasm of potential imitators.

The shortcomings of Penck's and Volz's methods, namely the failure to show clear language boundaries, led Geisler (1926) to attempt a geographical modification of the choropleth method. Instead of using administrative districts to represent the percentage figures of the different language groups, which had been done in the Polish maps, he advocated natural landscape units, such as lowlands, coastal lands, ground moraine lands, beet-growing areas, and plowlands. As an example of the necessity to observe these units he cited the Vistula lowland region, whose communities formed a partnership for survival in their joint task of dike management. After the division of the German-Polish region into landscape units, the percentage of Germans in these subdivisions was determined on the basis of the Prussian census of 1910. The results showed the dominance of Germans in the Vistula region and three bridges of German domination between Pomerania and East Prussia (see figure 3.2).[96]

Geisler emphasized that political boundaries had to respect the borders of natural landscape units which could not endure dismemberment. His geographical method was obviously inspired by the geo-organic concept of national territory, which also argued for the need to preserve the integrity of geographical regions, albeit on the scale of subdivisions of states (see Chapter 4). Geisler believed that his method yielded objective results, for it was based on unequivocal scientific criteria. However, he cautioned that

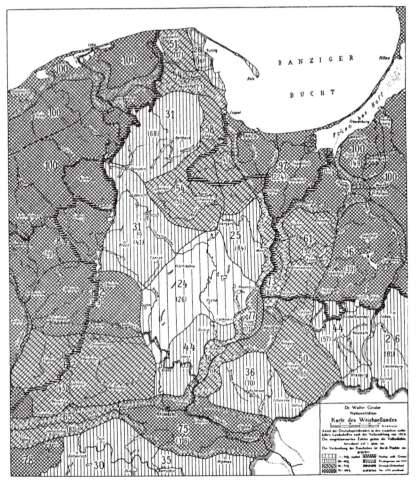

Figure 3.2 The distribution of national groups in the corridor area based on Walter Geisler's "geographical" method (Geisler 1926).
The map depicts the quota of German speakers in the individual natural landscape units based on the 1910 census. The figures in brackets indicate population densities per square km. The distribution of Kaschubes is indicated by dots. The categories of the legend are (from top to bottom): LEFT: 1–24%; 25–49%; 50–74%; 75–100%; RIGHT: present-day political boundaries; boundaries of the Empire in 1914; territory of the German Order; Prussian before 1772.

objectivity could only be ensured if this method was applied in a consistent manner, and he urged that this be done even if it proved unfavorable to the German cause.[97]

Geisler outlined his method very sketchily and it appears inconsistent. For example, he used natural landscape elements, such as coastal lands, in combination with agricultural structures, such as plowlands. While the former is a classification that is independent of human activity, the latter is a direct outcome of it. The interchangeable use of these two types of units introduced a bias, since the distribution of the nationalities is directly related to the quality of agricultural land, but not necessarily to other physical features of the environment. Historically, the Germans as the colonizing and governing nationality, especially during the territorial expansion of the Teutonic Order, were concentrated on the best agricultural land. In a different context, this distribution pattern was also pointed out by Trampler (1934a: 23).

Using elements from both Penck's and Geisler's methods, Milleker (1937: 642–5) developed an approach that would portray geographical spheres of influence of the different language groups based on aspects such as landed property, cultivation forms, and ownership of the means of production.[98] He wanted the maps, which he called ethnographic land registers (*Kataster*), to be at extremely large scales so as to be able to show the absolute figures for each group. As a result, this type of map would not only reveal detailed information on the distribution of language groups, but would also show which group was culturally more advanced.

Milleker's proposal had severe shortcomings. His call for the representation of each person at his or her home was completely unrealistic. The map would have been too large to allow an overview and its costs certainly would have been prohibitive. It is not surprising, therefore, that he did not include a sample of a section of such a map to illustrate his argument.

MAPS AND THE UPPER SILESIAN PLEBISCITE

The issue of map falsifications did not come to the fore in Germany just as a response to the peace conference, but also during the plebiscite in Upper Silesia.[99] As Stahlberg pointed out:

> It is not the intention of the Polish-French propaganda to reveal the truth, but to falsify the actual situation in the Polish favor. The cartographic representation of the plebiscite results again acted in this service. Namely, the falsification was based on the same principles as in the case of the language maps. What proved successful in that case was again applied here.
>
> (Stahlberg 1921: 12)

Again, the Spett map provided an impulse to German cartographic activity. On the day of the plebiscite, 20 March 1921, the French periodical *Le Temps* published a section of the infamous map as evidence of the Polish character of the area. About two months later, Albrecht Penck responded cartographically with a triptych of maps in the *Deutsche Allgemeine Zeitung* (Penck 1921b). It contrasted the section of the Spett map in *Le Temps* with two maps he made of the plebiscite results. One of the plebiscite maps used the choropleth method to depict the percentages of Germans and Poles based on districts (*Kreise*), the other presented the results using Penck's new method.

Penck (1921b) had two intentions behind the map comparison. One was specific: he wanted to show that the Spett map was blatantly wrong. Both of his plebiscite maps succeeded in this regard since both presented Upper Silesia as predominantly German. The other intention was broader: he sought to reveal the general shortcomings of cartographic representations based on districts, that is, choropleth maps. It was for this purpose that he included two different map designs for the plebiscite results. In his view, only the map using his modified version of a dot map, that is, his absolute representation method, was "cartographically and statistically perfect," because it considered population densities.

There were good reasons for Penck to draw public attention to the superiority of the map employing his graduated version of a dot map (i.e. his absolute representation method). Even though the choropleth map showed that most districts in Upper Silesia were German, it still supported the cession of the Polish majority districts of Ribnik, Pless, and Tarnowitz. Only the intermixed distribution in the graduated dot map stressed the indivisibility of Upper Silesia and therefore its wholesale retention by Germany. In addition, unaware of the crusade of German scientists to expose cartographic misrepresentations, most German newspapers had published a choropleth map to report on the plebiscite results (Stahlberg 1921: 18).

After the plebiscite results were in, the Poles apparently submitted a new map to support the rightfulness of their claims. This cartographic depiction of the results clearly showed a German western part and a Polish eastern part in Upper Silesia (Stahlberg 1921: 15). German scientists vehemently rejected this map as being a lie (Stahlberg 1921; Eckert 1925: 146).

This Polish plebiscite map, which I have so far not located, was described by Stahlberg (1921) in detail. It showed the percentages of Germans and Poles in the following way: each locality (*Gemeinde*) with more than 50 per cent Polish votes was given a Polish symbol and each locality with more than 50 per cent German votes received a German symbol. According to Stahlberg this was unjust: a locality with 40,000 inhabitants counted as much as a locality with 40. Since the Germans were concentrated in the larger settlements the method clearly disadvantaged the Germans. He gave an example of one case in which the German majority was achieved by a vote of 44,022 German votes against 8,588 Polish votes and compared it to a Polish majority which was the result of 20 Polish votes against 14 German votes. Yet, both carried the same graphic weight. In the view of Max Eckert, the use of this distorting method to support Polish claims compromised the scientific character of the map and turned it into a "prostitute" (Eckert 1925: 146).

The plebiscite results in Upper Silesia were also represented in large-scale German maps. Two maps used a method of representation in which each polling place was represented by circles of varying size to indicate the number of inhabitants. The circles were divided into colored sectors to show the quotas for the different votes. This method was an adaptation of the symbols for towns in Albrecht Penck's small-scale map in the *Zeitschrift der Gesellschaft für Erdkunde zu Berlin* (Penck 1919c).

One of the large-scale plebiscite maps was at a scale of 1:100,000 and prepared by the *Oberbergamtsmarkscheiderei* (mine-surveying institution) in Breslau, the other, at 1:200,000, was the work of the cartographic division of the *Preußische Landesaufnahme* (Prussian Topographic Survey).[100] While these two maps were considered accurate, a map by the *Preußisches Statistisches Landesamt*, at 1:100,000,[101] was heralded as the best and most accurate cartographic representation of the plebiscite results (Stahlberg 1921: 21–2; Eckert 1925: 477). It emulated Albrecht Penck's graduated dot map method, which he had first used in his 1:100,000 map in 1919.[102] Stahlberg had high hopes for this map:

> With this map, every legitimate demand on a cartographic representation of the plebiscite results is actually satisfied, and at the same time, a model was created based on which one would also like to see several other national questions of our border regions to be treated.
>
> (Stahlberg 1921: 22)

> Such a map of the Upper Silesian plebiscite belongs in every one of our schools, from the most elementary to the most advanced. . . . It is up to us now to make the plebiscite maps of Upper Silesia into the educational tool for national identity that they have the potential to be.
>
> (Stahlberg 1921: 26)

Apart from this general appeal to use the Penck-inspired plebiscite maps in nationalist education, Stahlberg also stressed the necessity of getting the German government to use this type of map during the upcoming Allied negotiations on the final boundaries of Upper Silesia: "The German government has to get these German trumps to the right authorities and into the right hands so that they will bear witness to the reality and truthfulness of the German right to Upper Silesia" (Stahlberg 1921: 26).

This was precisely the intention of the Foreign Office, which – in marked contrast to its neglect of maps during the peace conference – finally recognized the usefulness of maps to argue for the indivisibility of Upper Silesia. Two days before the plebiscite in Upper Silesia, a memorandum by a staff member of the Foreign Office stated that it would be most desirable if the plebiscite votes occurred in an intermixed distribution of pro-Polish and pro-German votes, rather than in large contiguous regions with pro-Polish votes. In case the results showed this intermixed distribution, the memorandum suggested that color maps be drawn which would reveal this pattern and thereby illustrate the indivisibility of Upper Silesia. The maps were to be employed in propaganda efforts abroad.[103] The suggestion was taken up: the plebiscite map by *Plankammerinspektor* Gerke of the *Preußisches Statistisches Landesamt*, which was in color and which showed such an intermixed distribution as a result of the absolute method it employed, was made at the request of the Foreign Office. However, there is no record of the use of the map in official propaganda abroad. It appears that the Foreign Office's active interest in maps was short-lived. Subsequent involvement of the Foreign Office in the mapping of German national territory was always in response to outside initiatives.

German scientists' heightened concern for accuracy in the mapping of national issues, which dominated the years immediately following the presentation of the peace conditions in May 1919, had important repercussions. It resulted in a greater understanding of the limitations of maps of national territory and in the development of new cartographic methods. This already proved useful in cartographic depictions of the plebiscite results in Upper Silesia. Various institutions adopted the novel techniques of German geographers, which enabled them to argue more effectively for the indivisibility of Upper Silesia.

In general, the production of more accurate maps gave scientific backing to the Germans' belief that injustice had been done to them in the Treaty of Versailles. This belief was confirmed after the plebiscite in Upper Silesia. Even though German scientific evidence – that is, maps with novel forms of representation – showed that Upper Silesia was indivisible, the Interallied Commission decided to partition the plebiscite area. As a result, revisionism appeared as a just cause, and scientists were motivated to put their research and expertise at the service of restoring the German nation.

4

NEW CONCEPTS OF NATIONAL TERRITORY

Whoever opens, for example, a German atlas from the pre-war period today, with its unreflective equation of language and nationality, will have a striking realization and he will say to himself: it could not have been any other way than that all these seemingly objective and in reality untrue – because they were undifferentiated – representations of the ethnographic distribution directly provided our enemies with the means against Germany: they showed them how Germany should be torn apart.

(Karl C. von Loesch, November 1921)[104]

The cry of injustice, which was the immediate German response to the peace treaties, received scientific backing through the production of more accurate language-based ethnographic maps. This new group of maps developed out of the German research on falsifications in foreign maps. Even though they allowed further-reaching claims than the existing maps, they did not suffice to reclaim all the lost territories. As a result, German scientists searched for alternative concepts which used a variety of data to explain why more territory should belong to Germany.

The plebiscite in Upper Silesia pointed toward a possible solution. The results revealed that language and nationality were not synonymous and that this discrepancy favored the German cause: 300,000 Polish-speaking persons had voted to remain with Germany, compared to 5,000 German-speaking persons who had opted for Poland (W. Volz 1922: 223). Provided that this plebiscite result was presented in an effective manner, it would prove that all of Upper Silesia was German (Stahlberg 1921: 20), a case which could not be made using language maps.

German scientists were convinced that *Sprachgemeinschaft* (language community) and nation, which coincided in Western Europe (for example, along the French–German border), varied considerably in Eastern Europe. The German geographer Otto Maull (1925: 412–16) argued that it was therefore necessary to draw maps of nations (*Nationenkarten*). He admitted that the data for such maps were difficult to come by, but that they could

49

Material zu einer Nationenkarte.

Figure 4.1 Otto Maull's map of nations (Maull 1925: 415).

be obtained in some ways, such as through voting patterns for national parliaments. He presented a map of the disputed Polish corridor based on votes for the Polish party in 1903 and 1871 (see figure 4.1). Although the map appeared to substantiate the predominantly Polish character of the corridor, Maull believed it showed that the Kaschubian northern part was a separate entity from the southern Polish part (Maull 1925: 415). Since many Germans were convinced that the Kaschubes were different from the Poles, Maull's map could be used to reclaim the coastal area between East Prussia and the German mainland. However, the large Polish area to the south, principally the district of Poznan, was lost, which was unacceptable to German nationalists. This was partly the result of Maull's choice of the choropleth method, which obfuscated the intermixed distribution of nationalities in Eastern Europe. Had Maull opted for a dot map, the Polish majority would not have been quite as obvious. Yet, it was not possible to claim the area as overwhelmingly German.

EMERGENCE OF SPATIAL DEFINITIONS

Since language or national identity, that is, the character of the population, proved insufficient to reclaim the lost territories, German geographers focused their attention on the German character of the land itself. Social concepts of national territory were replaced by spatial or geographical definitions of national territory. In contrast to language and nationality maps, which could be produced by a variety of other academic disciplines, such as ethnography, history, law, and statistics, the new spatial definitions were grounded in geographical and geopolitical theory, and were clearly the domain of geographers. They had great potential for far-reaching territorial demands and established the preeminence of geography and geopolitics in the fight for a revision of the Paris treaties.

Maps outlining the historical extent of German territory also offered the possibility of far-reaching claims, but solely on the basis of historical precedence, not national self-determination. Since the Paris peace conference had made the latter into the guiding principle of the new political boundaries,[105] demands for revision only appeared promising if the maps proved that the lost territories were ethnically German. Therefore, historical maps were not a good alternative; but they were useful as additional evidence or to build morale at home by raising awareness about the large territories previously controlled by German rulers, such as the Holy Roman Empire.

The inspiration for the new geographical concepts of national territory came from political geography and a school of thought which became popular after the First World War, *Geopolitik*. At the root of *Geopolitik* and political geography was the philosophical assumption that the state is a living organism. This theory had been developed systematically by the German geographer Friedrich Ratzel (1897) and the Swedish political scientist Rudolf Kjellén (1917). Ratzel, who was the founder of modern political geography, posited that the internal cohesion of a state was the result of the relationship between the people and the land as expressed in agricultural use. This relationship was so potent that the one could not exist without the other: they formed a living organism. Since life was characterized by movement and growth, the same applied to the state: states strove to expand their territory. They attempted to overcome "inorganic" political boundaries and to occupy their natural regions, which could cover an entire continent. Kjellén's theory encompassed an extension of Ratzel's geographical determinism. Apart from the influence of environmental conditions, for which he coined the term *Geopolitik*, Kjellén included the population (*Demopolitik*), economy (*Wirtschaftspolitik*), society (*Soziopolitik*), and power (*Herrschaftspolitik*).[106]

Although Ratzel's and Kjellén's theories provided a strong justification for expansionism, their application to maps of German national territory was not a straightforward issue. A central tenet of the geographical determinism of Ratzel's *Politische Geographie* and Kjellén's *Geopolitik* was the demand to bring political boundaries into agreement with the natural environment. But German geographers realized that they could not make a convincing argument for "natural" borders in the east.[107] There was only one rather meek attempt: Hermann Lautensach's (1924) use of an ill-defined band of natural features, called *Warägischer Grenzsaum*. But the fact that the cultural limits he added – for example, the eastern limits of German language in commerce and of German law in the Middle Ages – did not coincide with the natural one made it not very convincing (see figure 4.2).

Appropriately modified, Ratzel's and Kjellén's theories proved very

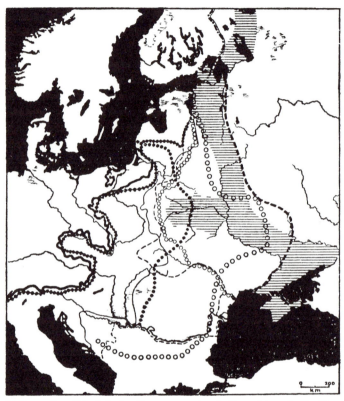

Abb. 2. Der Warägische Grenzsaum (nach Penck) und die Ostgrenzen der deutschen Kultur (nach Kaindl).

———— Ostgrenze des geschlossenen Verbreitungs-
gebietes des römisch-katholischen und evan-
gelischen Christentums.

▬▬▬ Ostgrenze des geschlossenen Verbreitungs-
gebietes der deutschen Sprache.

oo oo ooo Ostgrenze des amtlichen Gebrauchs des
Gregorianischen Kalenders.

o o o o Ostgrenze des mittelalterlichen Gebrauchs
des deutschen Rechtes.

▬ ▬ ▬ ▬ Ostgrenze deutscher Verkehrssprache im
Handel.

♦♦♦♦♦♦♦ Ostgrenze des geschlossenen Gebiets
deutscher Schriftart.

---------- Ostgrenze des geschlossenen Gebiets la-
teinischer Schriftart.

·—·—·—· Ostgrenze der Verbreitung beider zusam-
men.

•••••••• Ostgrenze fränkischer Hausformen.

Figure 4.2 Natural boundaries and limits of German culture in Eastern Europe
(Lautensach 1924: 472).

useful for claiming more national territory through maps. Influenced by economic, military, and *völkisch* considerations, German academics developed three new concepts which found cartographic expression: the geo-organic unity concept, the negative definition of territory, and the *Volks- und Kulturboden* concept.

The concept of geo-organic unity

The development of this concept was stimulated by the plebiscite in Upper Silesia. Being aware of the intermixed distribution of nationalities there, German scientists and officials knew that they could only argue for the wholesale retention of the district if they could show that the region was an indivisible unit. Just as in the case of ethnographic maps with novel forms of representation (see Chapter 3, "Maps and the Upper Silesian Plebiscite"), German geographers offered their expertise for this national cause.

Using the geopolitical idea of the state as an organism – in particular Kjellén's focus on population, economy, and society – they pointed to changes in the cultural landscape to support their claim that Upper Silesia was an integral and vital "organ" of the German state organism. Presented in this manner, the division of Upper Silesia not only promised dire consequences for the region itself, but threatened the very existence of the German state.

In 1920, that is, before the plebiscite of 20 March 1921 which was to decide the fate of Upper Silesia, Wilhelm Volz of the University of Breslau published a booklet with eight maps which confirmed the long-standing integrity of this region as well as its German character (W. Volz 1920). Apparently, Volz received government support to prepare the work, and it was so well regarded abroad that he was even made a member of the Oxford Geographical Society because of it.[108] After the plebiscite results for Upper Silesia were in, but before the final boundaries were set, Volz and a geographer from the Technische Hochschule Breslau, Bruno Dietrich, presented additional material to prove that Upper Silesia could not endure division and had to remain with Germany (W. Volz 1921a and 1921b; Dietrich 1921).

In his German-language work Volz (1921b) interpreted the plebiscite results on the basis of the cultural and economic development of the region. In the explanation to one of his maps,[109] he posited that since ancient times the economic and cultural make-up of Upper Silesia was dominated by the German element. He argued that the internal cohesion of the region was the result of German centers and paths of strength (*Kraftzentren und Kraftlinien*).

Volz's other work (1921a), which was in English and therefore solely directed at audiences abroad, attempted to prove that: "The whole of Upper Silesia, and particularly the entire industrial district, through its delicately interwoven mechanism becomes an indissoluble entity, from which no piece can be taken without far-reaching consequences" (W. Volz 1921b: 20–1).

His maps identified the interwoven mechanism as the system of railroad, water, and electrical lines which enabled the functioning of the burgeoning economy of Upper Silesia (see figure 4.3). He also made clear in a map and graph (W. Volz 1921a: 30, 33) that the Upper Silesian entity was only integrated into the German economy. The graph showed that the traffic of goods between Upper Silesia and Poland was negligible not only compared to the traffic of goods between Upper Silesia and Germany, but also compared to that between Upper Silesia and other neighboring countries. The map reinforced this message: it depicted the number of railroad lines in Upper Silesia and Poland to support his argument that Poland would not be able to cope with increased traffic demands because of an underdeveloped rail system.

Bruno Dietrich (1921), an expert in economic and transportation geo-

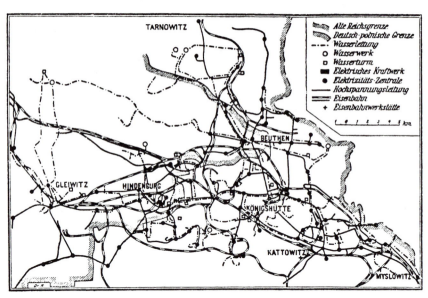

Eisenbahnen, Wasser= und elektrische Hochspannungsleitungen im oberschlesischen Industriegebiet.

Figure 4.3 The geo-organic unity of Upper Silesia as expressed in the interwoven system of railroad, water, and electric lines (Jäger 1928: 38).
This black-and-white map was included in two articles by the German geographer Fritz Jäger, which surveyed the effects of the new German–Polish boundary (Jäger 1924 and 1928). It is a reduced version of W. Volz's map (1921a: map 1). The only difference is that Jäger depicted the old and new German–Polish boundaries, while Volz only showed the old boundary.

graphy, argued in a similar manner. He offered three maps which showed that the German cultural boundary followed the pre-First World War political boundary. In his view, the maps proved that Upper Silesia showed different types and patterns of settlement and a more sophisticated road and rail system than the adjacent Polish areas. Thus it constituted a self-contained and well-functioning economic unit.

Volz's German-language work (1921b) as well as Dietrich's (1921) work were also intended for audiences abroad since their maps were trilingual (German, English, and French). Both were sent to the German government by their joint publisher, but they arrived too late to be presented at the meeting of the Interallied Commission which decided upon the final boundaries of Upper Silesia.[110]

Slightly modified, the geo-organic concept could also be applied at different scales. As we have seen in Chapter 3, Geisler incorporated the idea of small organic units into choropleth mapping of ethnographic data. Others, such as Langhans-Ratzeburg (1929), Haushofer (1927), and Trampler (1931: esp. 232–3) contrasted the political division of Europe before the war, which was deemed to be "organic" and more in agreement with geopolitical conditions, with "zones of friction or rupture" (*Reibungsgürtel* or *Schütterzonen*) after the peace settlement.

The geo-organic concept has typical elements of traditional geopolitical thought. The notion of "centers of strength" (*Kraftherzen*) is reminiscent of "pivots," "heartlands," or "core areas" (Mackinder 1904, 1919; Whittlesey 1939), and "zones of rupture" have reappeared as "shatterbelts" (Cohen 1991) or "*zones de tension*" (Lacoste 1986: 27). Yet, despite this emphasis on the influence of the environment, the geo-organic concept of national territory also contained hints of a decisive influence of culture. For example, Volz's stress on the ancient roots of German culture in Upper Silesia, which he succinctly stated as: "The forest is Polish; and the culture is German" (W. Volz 1921b: 10), clearly reveals his belief that the superiority of German culture played an important role in the creation of the organic unity.

The *Volks- und Kulturboden* concept

In the mid-1920s, a new and most powerful concept became popular in Germany: *Volks- und Kulturboden*. The first part of the term, *Volksboden*, had enjoyed wide currency before and was often used in maps to indicate the area inhabited by German-speaking people, that is, as a synonym for the German-language area (Meynen 1935: 108–9, 112–13). At the same time, the expression had radical *völkisch* connotations. The connection between the *Volk* and the soil (*Boden*) was a key element of *völkisch* ideology, particularly of the racist mainstream that was the precursor of National Socialism (Mosse 1981: 4–6, 15–16). The rootedness of the German *Volk* in the landscape provided the link to an idealized agrarian past. German peasants harmoniously united with nature formed a community with a "transcendental essence" (Mosse 1981: 4). Blood and soil defined the identity of the German *Volk*.

Neither the simple equation of *Volksboden* with the German-language area nor its *völkisch* interpretation made the term *Volksboden* in itself a useful concept with which to argue for more territory: much of Upper Silesia and the corridor was clearly made up of Polish-speaking areas. It took the brilliant idea of combining it with the term *Kulturboden* to create an entirely new definition of national territory. Albrecht Penck, who coined the composite term, has to be credited with this achievement.

Penck defined the *Volksboden* as the area where the German people settled, where the German language could be heard, and where the results of German work could be seen (Penck 1925: 62). The German *Volksboden* was accompanied by the German *Kulturboden*, which reached to the outer limits of the German culture area. In the West, the *Kulturboden* coincided with the *Volksboden*, whereas in the East it extended far beyond it. He pointed out that the German *Kulturboden* was characterized by an extremely careful and intensive cultivation, which unlike the land-use practices of other cultures did not halt where agriculture became difficult. All features of the cultural landscape showed this German character: the fields, the transport network, and the settlements (p. 64). German *Kulturboden* was "the result of German intelligence, German industriousness, and German labor" (p. 72). This made it possible to claim not only all lost territories, but even areas outside pre-war Germany simply by pointing to their German cultural character (see figure 4.4).

As in the case of the geo-organic concept, Ratzel's theory of the state as an organism offered inspiration. It not only supported a call for the indivisibility of organic regions, it could also be applied to the relationship between people and territory. Ratzel (1897) posited that the inner strength or cohesion of a state was a direct outcome of the intensity of land use: sedentary agriculture provided the most powerful bond between the population of a state and its territory, nomadism the weakest. Following this line of argument, the still visible imprint of the more intensive German type of land use in the area of the *Kulturboden* clearly showed that Germans had established the strongest bond with the area and thus had the right to claim it.

Although Penck's concept found great resonance, there was no universally accepted definition. Its intellectual and ideological origins were complex and allowed for different interpretations. In addition to *völkisch* ideas and Ratzel's theory of the state as organism, the concept was also affected by methodological discussions within geography. The 1920s witnessed serious disagreements about the study of landscape regions (*Landschaften*)

Karte des deutschen Volks- und Kulturbodens.

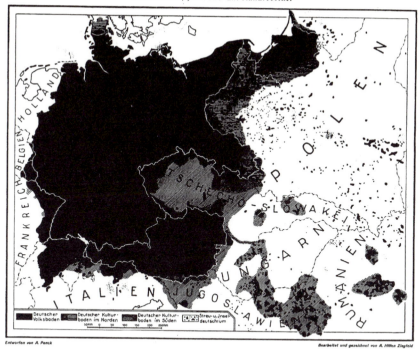

Figure 4.4 Albrecht Penck's concept of the *Volks- und Kulturboden* (Penck 1925: between pp. 72 and 73).
The four categories of the legend are (from left to right): 1) German *Volksboden*; 2) German *Kulturboden* in the north; 3) German *Kulturboden* in the south; 4) scattered German territory.

which directly impinged on the identification of the cultural landscape boundaries required when delimiting the *Volks- und Kulturboden*.

Differences in the interpretation of the German *Volks- und Kulturboden* become clear in a comparison of the most important proponents of the concept. Wilhelm Volz, who was involved in the first major research work which applied the concept to Germany's eastern borders (W. Volz 1926), expanded Penck's definition of *Volksboden*. In his view, the German *Volksboden* reached much further east and included parts of what Penck had classified as *Kulturboden*.[111] He argued that the non-German-speaking groups within the pre-war boundaries of Germany, such as the Kaschubes and Masures, were part of the German nation (*Volk*). The superior German culture had transformed these Slavic peoples: "The Kaschubes and Masures, the Upper Silesians and the Sorbs are German in culture, members of the German nation, Germans, even if the old idiom is not yet extinct; their national will and their national consciousness is German" (W. Volz 1926: 6).

As a result, the German *Volksboden* in the East included all these ethnic groups. In other words, Germany could claim most lost eastern territory because the superior German culture had imprinted itself not only on the landscape, but also on the Slavs who settled there.

The application of the concept as outlined by Volz was not without problems. It was fairly straightforward to identify changes in the cultural landscape to show the outer limits of the German culture area, such as changes in agricultural practices, building styles, and settlement patterns. But it was rather difficult to prove that the Slavic groups in the former eastern border areas of Germany really felt like Germans except for the areas where plebiscite results were available (i.e. East Prussia and Upper Silesia).

Volz's version has to be seen as an extension of his geo-organic conception of national territory which stressed environmental influences over ethnic or racial ones. In fact, Volz explicitly rejected race as a determinant: "It is not race which determines nationality (*Volkstum*), . . . but the will and the national consciousness" (W. Volz 1926: 6). Thus Volz identified very closely with the view of Ratzel, who believed that cohabitation in the same space was more important than ethnic or racial identity (D. Murphy 1992: 12–13, 17). Ratzel, like most German geographers in the German Empire, had been an adherent of the idea of a political nation (Faber 1982: 396–8).[112]

By contrast, Penck was more prone to follow *völkisch* views. He stressed that the *Kulturboden* was not the result of special environmental conditions. It could be encountered anywhere in the world: not only in Europe, but also near the tropics and close to the Polar circle (Penck 1925: 70). In short: "However different the conditions are under which it grew, it still appears uniform everywhere, it is always characterized by great care in cultivation, by the comfort and cleanliness of the dwelling, . . . by well-maintained roads" (Penck 1925: 70). In Penck's view, it was all up to people with a definite predisposition (*bestimmt veranlagte Menschen*) to alter nature according to their wishes (Penck 1925: 70), a clear advocacy of racial influences.

The concept of the *Volks- und Kulturboden* was also addressed outside geography. Max Hildebert Boehm, the director of the *Institut für Grenz- und Auslandsstudien* and of the *Deutschtumsseminar* at the *Hochschule für Politik*, Berlin, incorporated it in his book on the German nation (Boehm 1932).[113] In an extension of Penck's theory, he posited that the nation (*Volk*) was a spatial organism which not only established links to the soil through agricultural use, but also was "rooted" through private property in cities and through the control of natural resources, such as coal-mining. The area where these activities took place constituted the *Volksboden*, for which he preferred the term *Volkssiedelboden*. The impact of these activities had a fundamental effect on the landscape, which became imprinted with

the culture of the nation; this was the *Kulturboden*. When a nation abandoned an area, the *Kulturboden* remained as testimony to its previous
activities. This meant that *Kulturboden* outside the *Volksboden* was only a
historical relic, a much more limited interpretation than Penck's. However,
Boehm also extended the spatial configuration of a nation beyond the
confines of the *Volks- und Kulturboden* with the category of the *Volkswirkungsraum*. The *Volkswirkungsraum* was the area which was connected
to the nation (*Volk*) by its "dynamic" component, that is, commerce, and
penetrated into the *Volksboden* of neighboring nations. As examples of the
dynamic or mobile element of the *Volk* he mentioned traveling salesmen,
artisans, soldiers, and tourists.

Boehm (1932: 79) illustrated his categories with a biological analogy. The
rooted activities of the *Volk* created an "ethnic flora," that is, the *Volksboden*. But since the *Volk* was also fauna, it also roamed the land and
established connections, which made up the *Volkswirkungsraum*. Boehm
stressed that all activities of the *Volk* were in a sense dynamic and that
therefore the spatial extent of the different categories fluctuated. As a
result, Boehm was very critical of cartographic applications. He believed
that they led to fateful oversimplifications and that short of future innova-

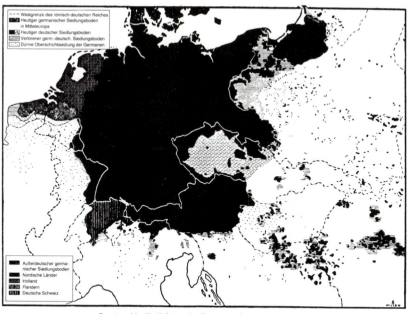

Der deutsche Volksboden in Vergangenheit und Gegenwart

Figure 4.5 German *Volksboden* according to Max Hildebert Boehm (Boehm 1930:
291). N.B.: Lettering of legends was slightly altered for legibility.

tions, such as cinematographic representations, they could not be achieved satisfactorily (Boehm 1932: 91). This explains why Boehm never published a map of the *Volkswirkungsraum*, which was the most dynamic component of his theory. However, two years before his major treatise, Boehm (1930: 291) had attempted a cartographic representation which gives an idea of what he considered to be the extent of German *Volksboden* and German *Kulturboden* (see figure 4.5). Although the map did not use the term *Kulturboden*, the horizontally shaded category "lost Germanic settlement area" (*verlorener germanisch-deutscher Siedlungsboden*) corresponds to Boehm's definition of *Kulturboden*. A comparison of this map with Penck's map (figure 4.4) reveals that Boehm claimed less *Kulturboden* in the southeast and also excluded Swiss German territory by giving it a separate category.

The negative definition of national territory

This concept did not stake a specific territorial claim, but rather argued that the existing territory was insufficient and therefore had to be expanded. It was inspired by Kjellén's stress on the embodiment of power in a state

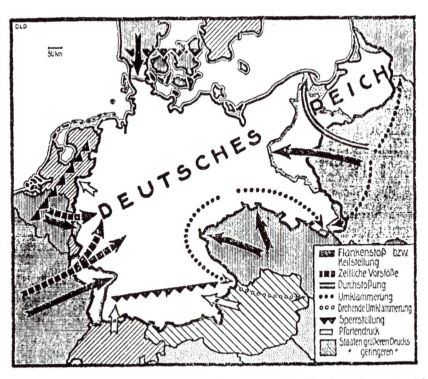

Figure 4.6 Germany's strategic shortcomings as a result of her irregular territorial shape I (Maull 1928: 333).

organism,[114] for which security was a prerequisite. Followers of German *Geopolitik*, who were its most ardent supporters, such as Ernst Tiessen (1924), argued that Germany's post-war boundaries violated the requirements of good state borders. The most advantageous boundaries for a state were deemed to be as short and straight as possible, that is, in its ideal form a circle, because this ensured easy defense (Tiessen 1924: 206).

Thus, the central assumption of the negative concept was that the existence of Germany was threatened because of the irregular shape of its territory. This argument was substantiated with maps which showed that the jagged course of the boundaries facilitated not only the isolation and dissection of exposed districts, such as Silesia and East Prussia, but also attacks on the center of German territory. Impending strangulations and pincer movements were illustrated with powerful clamp-like half-circles and piercing arrows (Springenschmid 1934: 10), the vulnerability of the center with arrows pointing out the short distance to the capital from the eastern

Figure 4.7 Germany's strategic shortcomings as a result of her irregular territorial shape II (Ströhle 1931: 75).

boundaries (see figures 4.6 and 4.7). The threat to the very survival of Germany was also given strong verbal images by calling the existing state "mutilated" (*verstümmelt*), a clear reference to the prevailing view in German geopolitics that the state was a living organism.

In the eyes of geopoliticians, the situation was aggravated by military restrictions imposed on Germany in the Treaty of Versailles, such as the prohibition of fortifications in border districts, the demilitarization of the Rhineland, and the abolition of the Air Force. This reduced the power base of the German state even further and thereby gave added validity to the demand for more territory and rearmament. Some of the maps shaded all areas within the German boundaries which had military restrictions: this left only a small triangle of sovereign territory in East Prussia and a much-truncated sovereign German mainland. Other maps showed the reach of foreign artillery (Haushofer and Trampler 1931: 93): the German state territory depicted in solid white and the projected impact area in solid black, the Germans were left only with a small area in the northwest and center, a hopeless situation.

Despite the emphasis on military power – a further confirmation of geography's long-standing interest in warfare – the concept also reflected *völkisch* views. Since much of the threat concerned the eastern border regions, the negative definition also supported the continuously invoked claim that the Slavs sought to push Germans out of Central and Eastern Europe. In addition, it gave credence to the belief that there was a general conspiracy against Germans; Germans were being victimized, which left them no choice but to band together as a community, to seek solace in the *Volk*.

POTENTIAL OF SPATIAL CONCEPTS FOR TERRITORIAL CLAIMS

The newly developed spatial concepts of national territory provided attractive alternatives to social definitions, such as language use or plebiscite votes, because generally they allowed for further-reaching claims. They also matched the anti-democratic stance of the nationalists who propagated them. National territory was defined by the landscape or the strategic situation; there was no need to give people a choice in an election.[115] However, some of the spatial concepts also had disadvantages. The geo-organic concept is a case in point. On the one hand, it offered a good argument for the return of all territories ceded in the Versailles Treaty – especially the portion of the heavily industrialized part of Upper Silesia which was given to Poland following the 1921 plebiscite. But on the other hand, it effectively contradicted claims to ethnic German territories which were outside the pre-war boundaries of Germany, such as Austria and the Sudetenland. These areas, albeit predominantly German, obviously had been part of the state organism of Austria-Hungary and therefore could be claimed as organic components by the successor states of the Austro-Hungarian Empire.

The negative definition did not outline a specific territory and thereby left claims very flexible and open-ended. At the same time, this prevented it from being a real alternative and relegated its use to supporting arguments for other concepts. Its main advantage for revisionist aspirations was that it made military build-up and expansion appear as necessities for the survival of Germany. Thus, the concept was helpful for the preparation of war.

The *Volks- und Kulturboden* concept clearly had the greatest potential. It laid claim to all ethnic German areas (i.e. the *Volksboden*) and to all areas which were part of pre-war Germany or which previously had been colonized by Germans (i.e. the *Kulturboden*). Claims could be extended even further by pointing to the *Volkswirkungsraum*, that is, the area where German cultural influence was dominant. Since the latter category, which had been developed by Boehm (1932), encompassed a vast region in the East, it nicely fitted the Nazi demands for *Lebensraum* there. In fact, Boehm's term for the combination of *Volks- und Kulturboden* and *Volkswirkungsraum* was *Volkslebensraum* (Boehm 1932: 105).

The close affinity of the concept of *Volks- und Kulturboden* with *völkisch* ideas, which acquired a mass political base in the Weimar Republic (Mosse 1981: 5), also gave it the potential to be widely adopted. Penck did not leave it to chance for this to happen and set out to popularize it with maps. For Penck, the concept entailed a mission since "the German *Kulturboden* is the greatest achievement of the German people. . . . A small number of Germans suffices to transform a large country into German *Kulturboden*" (Penck 1925: 69–70). Therefore, it was necessary to make it public knowledge:

When our youth is being taught from childhood about the cultural tasks our people has achieved, when at a more mature age it is being instructed about the German *Kulturboden*, when the latter is thoroughly studied at the universities, researched by our scholars, [and] cherished by all, then a feeling of strength will develop in the nation, which is not just intoxicated with its *gloire* and which is not just cheers of hurrah, but which is anchored in the soul of the people.

(Penck 1925: 72)

Penck's mission was a success. The concept of the *Volks- und Kulturboden* became enormously popular in the Weimar Republic: it was used extensively by geographers, it was a catchword among conservatives (Schultz 1990: 64), and it was incorporated in the curriculum of schools in Prussia and other regional states (Fahlbusch 1994: 162). Penck has to be given much of the credit for this. He publicized his concept with an extremely effective map, which was drawn by one of the masters of suggestive

63

cartography, Arnold Hillen Ziegfeld (see figure 4.4).[116] He even disseminated it in a slightly modified version as a wall map, which took inspiration
from Boehm's category of the *Volkswirkungsraum* and Lautensach's (1924)
map (see figure 4.2). The wall map showed that other aspects of German
culture reached far beyond the *Kulturboden* and incorporated into the
depiction the region where the German language was used in commerce.[117] Both of Penck's maps reached a wide audience because they
were reproduced in numerous books and school atlases.

Penck did not have a difficult time pointing out the advantages of his
concept for the German national cause. By showing great German achievements it gave Germans a renewed feeling of self-worth after the defeat in
the First World War.[118] It also provided scientific backing to the general
belief among German conservatives that Germans were culturally superior
and it supported tremendous territorial claims. Penck's *Volks- und Kulturboden* went far beyond pre-First World War Germany. It even included half
of Czechoslovakia and large tracts of land in southwestern Europe.

The concept even gave other research added validity. Historical research
on former German colonization as far back as prehistory found a new
purpose in the uncovering of ancient German *Kulturboden*. Studies of
Germans living abroad (*Auslandsdeutschtum*) were seen in a different light
since they were the outposts of the German *Kulturboden*: "Wherever Germans live communally and use the surface of the earth, it [the *Kulturboden*]
appears" (Penck 1925: 69). The currency of the term *Volks- und Kulturboden*
among academics and the general public was ensured when it was adopted
to identify the most important coordinating institution for representations
of German national territory in the Weimar Republic, the *Stiftung für Deutsche
Volks- und Kulturbodenforschung* in Leipzig.

5

COORDINATION OF CARTOGRAPHIC REVISIONISM
The *Stiftung für deutsche Volks- und Kulturbodenforschung*

The standardization of representation is the goal of scientific work.
(*Reichskanzlei* report 1929)[119]

New types of cartographic representation were developed in the aftermath of the Paris peace treaties to offer improved scientific arguments for their revision. One group of maps developed out of the German research on falsifications in foreign maps and led to the production of more accurate cartographic methods, such as Penck's version of the dot map. These maps were based on language or nationality (see Chapter 3). The other major group of maps focused on alternative concepts of national territory and was guided by the premise that neither language- nor nationality-based maps would suffice to reclaim all lost territories. Instead, these maps chose spatial concepts of national territory, such as the geo-organic concept, the negative definition of national territory, and the concept of *Volks- und Kulturboden* (see Chapter 4).

Both main groups of new maps were not just the work of individuals, but were early on incorporated into an institutional structure, which coordinated and supported their development, the *Stiftung für deutsche Volks- und Kulturbodenforschung*.[120] The *Stiftung's* influence was considerable: it organized meetings of scholars from various disciplines, provided support for research projects, funded publications, carried out research for the government, consulted government institutions, and collected data for a major reference work.

FIRST INITIATIVES

The *Stiftung's* early history is rather involved and it is difficult to identify its true beginnings and founding fathers.[121] The files at the German Foreign Office reveal that first initiatives go back to the end of 1921, when the Foreign Office received proposals by Karl C. von Loesch and Wilhelm Volz.[122] The proposals were very similar: they both represented the attempt to create an institutional structure which encouraged the production and dissemination of improved scientific arguments for a revision of the treaties, in particular maps.

The provenance of the proposals is revealing. One was authored by the leader of a major *völkisch* organization (Loesch), the other by a geographer (Volz). *Völkisch* organizations and geographers had similar views with regard to the representation of German national territory in maps. They cooperated not only in research and development, but also in the enforcement of a unified cartographic message (see Chapter 7).

Karl C. von Loesch, a member of the neo-conservative circle in Berlin who headed the *Deutscher Schutzbund*,[123] argued in his proposal for the creation of an institution which would unify and direct scientific research to help in the revision of the treaties. The goal was to show that the Paris peace treaties were based on false material – in particular cartographic representations – and to present improved versions.[124] The false material included not only foreign maps but also domestic cartographic products whose shortcoming was mainly to equate language and nationality. German science had the task of contradicting these maps with more accurate and more sophisticated methods of representation. Loesch identified several steps by which to achieve this:

1 Private congresses (no public access) for the exchange of ideas among scientists and politicians.
2 Financial support for German scientists to be able to conduct research.
3 Presentation of the research results in books, graphs, and maps.
4 Supply of foreign scientists or writers who are willing to collaborate with material and money.
5 Popularization of these works in Germany and abroad.[125]

He pointed out that this should not entail the creation of new research institutes, but rather the formation of a coordinating organization, which he called *Austauschzentrale*. Loesch suggested that any further details should be worked out in a meeting of scientists and political activists in early January 1922, for which he requested financial support.

The proposal by Wilhelm Volz outlined intentions similar to those of Loesch's *Austauschzentrale*. Volz requested financial support for the work of the newly created *Schlesische Gesellschaft für Erdkunde* in Breslau. The *Schlesische Gesellschaft* – which had been founded by Volz, the local chapter of the *Deutscher Schutzbund*, as well as Silesian government representatives and industrialists – was dedicated to scientific research on Germans in the East. Its main goal was a careful preparation of the revision of the Paris peace treaties: "so that in the hour of justice, which also must come for Germany, we stand prepared and do not have to compile the material again in a frenzy and thereby lose valuable time."[126] This required the following tasks:

1 Collection of all available material on Germans in the East.
2 Evaluating and supplementing this material.
3 Systematic scientific presentation of this material in written and cartographic form.
4 Suitable publications.
5 Study and examination of anti-German works.[127]

As in the case of the *Austauschzentrale*, the material in this context meant
mainly maps: the only examples which Volz and reports relating to the
Schlesische Gesellschaft gave were either the use of misleading German maps in
foreign propaganda, such as *Andrees Handatlas*, or the production of
improved maps, such as Volz's cartographic work on Upper Silesia.[128]

Because both proposals seemed to have the same intentions, but argued
for different solutions, the Foreign Office was not sure which one it should
favor. What further complicated the issue was that not just the *Deutscher
Schutzbund* was involved in both, but also Volz: Loesch mentioned that his
proposal was the collaborative effort of himself, Penck and Volz. After a
joint meeting with Loesch and Volz, the Foreign Office gave each of them
money.[129]

It is interesting to note that Volz had submitted a similar proposal to the
German government earlier without success. In summer 1921, he had
asked the Ministry for Science, Culture, and Public Education to support
the establishment of an institute for the study of cultural and economic
geography in Upper Silesia (Fahlbusch 1994: 56–7). The Foreign Office was
also made aware of this plan and it is surprising that the response was
negative given the time the proposal was submitted: July 1921, that is, after
the plebiscite results were in, but before the decision on the final boundary
of Upper Silesia. Fiscal constraints could have played a role (Fahlbusch
1994: 57), but the government might also have believed that an effective
presentation of the plebiscite results would suffice to retain the territory
(see Chapter 3).

It seems that Volz had created the *Schlesische Gesellschaft* to counteract the
influence of Loesch's *Austauschzentrale* and to secure his control of research
on Germany's eastern border regions, especially Upper Silesia. In a speech
held at a confidential meeting in Breslau on 23 August 1921, Volz stated
that plans were under way to create a central organization in Berlin which
would coordinate scientific efforts toward a revision of the Paris treaties.
He argued that:

Objections against this centralization are to be made especially in regard
to Silesia and the extraordinary amount of work which has to be carried
out here. Hence the demand: an independent institution for Silesia with
a regionally delimited scope . . . incorporated into the structure of the
aforementioned Berlin organization.[130]

He suggested that the Silesian institution be housed in the Breslau Institute of Geography, of which he was the director, because much work on the subject had been completed there. In support of his argument he pointed to the samples of such work which were exhibited at the meeting.

The first meeting to establish the *Austauschzentrale* took place at the *Gesellschaft für Erdkunde zu Berlin*, from 3 to 5 January 1922. Of the thirty-eight participants a quarter represented political groups or the government; the rest were academics, with geography being the single most important group with thirteen representatives, followed by history with five. The scientists came from various places in Germany and Austria, one of them even from Prague, Czechoslovakia, and they had been invited by the organizers (Loesch, Penck, and Volz) because they had done work which aimed at a revision of the treaties. It is therefore not surprising that they declared their willingness to place their academic work at the service of the German state and nation (*Staat und Volk*).[131]

The initial presentations at the meeting by Penck and Loesch outlined the general issues that required the founding of an *Austauschzentrale* from the viewpoints of science and politics, respectively. They were followed by lectures on regional cartographic works related to the Paris treaties: the newly developed cartographic methods of Penck and Volz, the use of maps at Trianon, Austrian responses to demands by southern Slavs, Sudeten-German research, as well as scientific works on the German–Danish, the German–Belgian, and the German–French border areas.

This sparked general discussions on methods of cartographic representation and their political effects. The participants agreed that in the future, the German settlement and language area should always be depicted with the most visible color: bright red. In addition, the development of cartographic representations using black and white was recommended. This was a reflection of the general post-war trend in Germany to pay closer attention to persuasive cartographic techniques (see Chapter 6).

Another big discussion at the January 1922 meeting addressed the shortcomings of areal coloring and statistical data collection. It resulted in two demands: first, that scientists either draw their own maps or at least closely supervise the execution of their drawing by draftspersons and at the printer, and, second, that, following the American model, the next German census should only contain a question asking about the ability to speak German and not about the mother tongue (*Muttersprache*). The dissemination of these demands and recommendations was entrusted to the *Austauschzentrale*, which was formed on the last day of the meeting and named provisionally "*Mittelstelle für zwischeneuropäische Fragen*." The *Mittelstelle* consisted of three academics (Albrecht Penck, Wilhelm Volz, and Joseph Partsch)[132] as well as two members of the *Deutscher Schutzbund* (Karl C. von Loesch and Albert Wacker).

The general function of the *Mittelstelle* was to be a central mediator and coordinator that "supplied the intellectual weapons of science to politics."[133] Its work, which was not to be made public, was outlined by the meeting as follows:

1 Coordination and exchange of information among scientists (German and foreign) and political organizations.
2 Allotment and delimitation of fields of study.
3 Development of universal terminology and methods of representation.
4 Repudiation of foreign and domestic works which are damaging to the German cause.
5 Support of German scientific works (production, dissemination in Germany and abroad, political application).
6 Encouragement of new works which are politically desirable.[134]

The list revealed that the *Mittelstelle* was to have even more overlap with the *Schlesische Gesellschaft* than Loesch's original conception of the *Austauschzentrale*. This was compounded by the fact that Volz managed them both in the first years of their joint existence: he was the head of the *Schlesische Gesellschaft* and director of the *Mittelstelle* in 1922 and 1923. During that time, Volz used the letterhead of the *Schlesische Gesellschaft* with the typewritten addition *"Mittelstelle für zwischeneuropäische Fragen."* When Volz became a professor at the University of Leipzig in late 1922, the importance of the *Schlesische Gesellschaft* faded. It had lost one of its most influential supporters, who now concentrated his efforts to expand the role of the *Mittelstelle*. The *Schlesische Gesellschaft* had served Volz well: it enabled him to secure considerable financial support for his own work since most of the successful requests for assistance which were not part of the regular budget were for Volz's publications.

SCIENTIFIC-POLITICAL CONGRESSES

The most important activity of the *Mittelstelle* was the organization of scientific-political congresses, for which it received financial support not only from the Foreign Office but also from the Interior Ministry.[135] Between 1922 and 1931, when the successor to the *Mittelstelle* was abolished, eighteen meetings were held in various places in the border regions all around Germany. The first two congresses which followed the founding meeting in Berlin were held in Marktredwitz (September 1923) and in Witzenhausen/Kassel (March 1924). These two were the most general in scope and devoted to the eastern and western German border territories. The remaining congresses focused on specific regional problems, such as Bautzen (September 1924) on the Sorbian issue, Heppenheim (October 1924) on the Rhineland, Lübeck (March 1925) on the German–Danish border, Marienburg (October 1925) on the Polish corridor, and Ortels-burg/East Prussia (September 1927) on Lithuania.

The meetings were confidential and attended on average by about forty to seventy participants. Apart from a few members of the *Deutscher Schutzbund*, government officials from the Interior Ministry, the Foreign Office, and regional institutions, the majority were scientists from Germany and Austria. At some meetings, scientists from Switzerland, Belgium, Czechoslovakia, the Memel territory, and Lithuania participated and at times even used aliases to protect their identity.

Just as at the initial meeting in Berlin, where important recommendations on map design were agreed upon, the congresses produced tangible results. They enabled scientists interested in developing scientific arguments for a revision of the treaties to exchange ideas, to find out about new projects, to agree on terminology as well as methodologies, and to receive input on political needs. At the 1924 meetings in Witzenhausen and Heppenheim, Albrecht Penck pointed out that the notion of German *Volksboden* should include cultural aspects and argued for the production of maps of German culture. This seems to have been the inspiration for his article on the *Volks- und Kulturboden* (Penck 1925). It also led to the change in the *Mittelstelle*'s name to "*Mittelstelle für deutsche Volks- und Kulturbodenforschung*" in late 1924.

The examination of the Sorbian issue in Bautzen (25–7 September 1924) sparked several initiatives: the plan for a publication to contradict two existing German works which were considered pro-Sorb; the idea of destroying the only existing map of the distribution of the Sorbs by Dr Muke, which was being sold at a bookstore in Bautzen and which was considered harmful to German interests;[136] and the production of a substitute map which showed the dominance of Germans in the area.[137] Volz and Loesch took up these ideas.

Barely three weeks after the meeting, the Interior Ministry informed the Foreign Office that the *Deutscher Schutzbund*, which was headed by Loesch, had taken action in regard to the Muke map: it had made sure that the Muke map was withdrawn by the German booksellers and that none of the remaining copies of the map were sold at the bookstore in Bautzen.[138] Volz saw to it that a "suitable" substitute was published as soon as possible: a new language map of the Sorbian area using Volz's method appeared in the 1925 issue of the *Mitteilungen der Gesellschaft für Erdkunde zu Leipzig*.[139]

At the congress in Emden (March 1927), which was devoted to Lower Germany, German scientists agreed that in the future Frisian was to be designated simply as a German dialect, and not as a separate language, in order to prevent further claims to German territory. This meant that the practice of the Prussian Statistical Office of identifying Frisians as a linguistic minority had to be changed. German linguists, in particular the eminent Friedrich Panzer from the University of Heidelberg (the father of

70

the geographer Wolfgang Panzer), declared their willingness to develop the scientific evidence to support this initiative.

Discussions at the congress in Ortelsburg led to the clarification of methodological aspects of language and nationality maps. It was agreed that for the eastern areas language maps needed to be supplemented by nationality maps and that their representational methods had to be different for detail and overview maps as well as for regular maps and wall maps. Other meetings drew attention to the danger of foreign activities: a speaker at the Lübeck congress pointed out the continued use of Danish maps which extended the Danish language area too far south and Penck stressed in a discussion at the Marienburg meeting that there was a need to resist foreign renderings of German place names.[140]

EXPANSION OF ACTIVITIES

By the mid-1920s, the *Mittelstelle* had consolidated itself. The conferences were held regularly and it became actively involved in research projects, publications, and consulting on map design. The *Mittelstelle* also had to expand its staff and hired the geographer Friedrich Metz, who had been appointed as a lecturer (*Privatdozent*) at the Geographical Institute in Leipzig. The financial needs of the *Mittelstelle* grew correspondingly, which prompted the Interior Ministry in June 1925 to suggest to the other main sponsor, the Foreign Office, that the *Mittelstelle* be converted into the legal framework of a foundation (*Stiftung*) to allow for "more responsible accounting."[141] The foundation was established on 30 October 1926 and the name of the *Mittelstelle* changed to "*Stiftung für deutsche Volks- und Kulturbodenforschung.*"

Although the work of the *Stiftung* was not intended to include active research, it was soon entrusted with a massive government project at the behest of the Interior Ministry: the scientific investigation into the negative effects of the post-Versailles boundaries (*Wissenschaftliche Bearbeitung der Grenzzerreissungsschäden*). The findings, which were to be used as the basis for a concerted government amelioration program, were also to be presented to the general public in the form of persuasive maps.[142] The area of the corridor, which in the eyes of the Interior Ministry was a "bleeding border" (*blutende Grenze*), was to be given primary consideration. The project entailed an official endorsement of the geo-organic unity concept, for it attempted to show that the new boundaries had severely damaged the health of the economic organism (*Wirtschaftsorganismus*) of Germany (J. Volz 1930/1: 39–40).

The initiative for this project appears to have originated in the *Stiftung*: during a meeting of the *Stiftung*'s advisory board on 1 December 1926, it was pointed out that a recent article by Fritz Jäger,[143] which dealt with the damaging effects of the eastern boundary provisions of the Versailles treaty, should be expanded upon by a series of large-scale suggestive maps. There was agreement that such a cartographic work would provide a convincing argument for government action and it was decided that the *Stiftung* would take care of it.[144]

With the endorsement of a variety of government agencies[145] the *Stiftung* collected an enormous amount of data through field trips and questionnaires, which were sent not only to all eastern border communities but also to numerous private and official associations and institutions (J. Volz 1930/1). A preliminary compilation of the material was submitted to the Ministry of Finance for use at the reparation conference in Paris, which was held between 11 February and 7 June 1929.[146] The final report of the project, which was completed at the end of 1929 by Wilhelm Volz and Hans Schwalm[147] (Volz and Schwalm 1929), was printed as a short memorandum and distributed among government agencies.[148]

The original plan of preparing a popular edition was not carried out. Instead, two of the maps from the memorandum which depicted the damaging effects of the new boundaries on the economy[149] were appended to a description of the project in a scientific journal (J. Volz 1930/1). Other material on the negative effects of the new boundaries, which had been compiled by the *Stiftung*, was published by a government commission in charge of investigating German economic conditions in 1930, that is, during the world economic recession (*Stiftung für deutsche . . . 1930*). One of the maps in the government publications depicted how a vital transportation road was dissected to award a hospital to Poland. The accompanying text explained that Poland did not need the hospital and that a new road connection was extremely expensive because of unstable terrain. Another map showed how the agricultural fields of a community were dissected by the new boundaries.

In addition to active research on *Grenzzerreissungsschäden*, i.e. the geo-organic unity concept, the *Stiftung* was especially involved in research dealing with the German *Volks- und Kulturboden*. It compiled and edited an ambitious reference work on Germans living in border areas and abroad, entitled *Handwörterbuch des Grenz- und Auslanddeutschtums,*[150] and it initiated, coordinated, and funded numerous mapping projects on German settlements in border regions and in the neighboring countries of Germany. Among them were regional atlases of Saxony and Bavaria,[151] Albin Oberschall's map of the nationalities of Czechoslovakia (1927),[152] Jakob Bleyer's map of the German settlements in Hungary (1928),[153] and a map project by Rudolf Spek on German settlements in Greater Romania.[154] The *Stiftung* also had connections with scholars involved in the production of the *Atlas der deutschen Volkskunde*, a monumental enterprise designed to catalog the morphology of German culture, such as dialects and rituals.[155]

Through these projects and the annual conferences, the *Stiftung* established an extensive network of connections which included leading scholars interested in German nationalism, regional research centers, and interest groups. For example, the *Stiftung* collaborated with the *Deutsches Ausland-Institut* in Stuttgart in the publication of a work on the distribution of Germans in Hungary (Rieth 1927), which included four maps. These contacts were not just restricted to Germany, but also extended to foreign countries. It collaborated with scientists, such as the Austrian scholar Wutte from Klagenfurt and the Swiss scholar Ammann from Aarau, with German national groups, such as the *Deutschpolitisches Arbeitsamt Prag*, which was the central institution of all Sudeten-German parties, as well as with research institutes, such as the *Historisch-Statistische Verbindungsstelle Bromberg*.[156] It even planned the founding of a new research institute at the University of Szegedin, Hungary.[157]

Despite these varied research activities, the *Stiftung* did not lose sight of its role as a mediator between science and politics. It supplied the *Reichs-wehrministerium* with maps and literature for disarmament negotiations in Geneva.[158] Drawing on the results of the Bautzen conference, it advised the Prussian Statistical Office (*Preußisches Statistisches Landesamt*) and the Census Office (*Statistisches Reichsamt*) on the inclusion of questions in census surveys regarding national minorities, such as the Sorbs.[159] It also consulted with publishing houses on the political appropriateness of map designs (see Chapter 7). One member of the Interior Ministry, Ministerialrat Tiedje, believed that the *Stiftung*'s expertise in cartographic representation should be exploited further to achieve greater uniformity in scientific works.[160]

In 1930, the *Stiftung* reached another milestone. After several proposals by Volz, the *Stiftung* finally started its own journal, the *Deutsche Hefte für Volks- und Kulturbodenforschung*. Now, the research inspired by the *Stiftung* had a forum and could be presented in a unified manner. However, the apparent success of the *Stiftung* was misleading. With the increasing importance of the *Stiftung* since the mid-1920s came internal dissent and power struggles.

Wilhelm Volz, the managing director, was the driving force behind amassing more power to the *Stiftung* and in particular to the main office in Leipzig. He wanted to make the *Stiftung* a purely scientific institution and argued for the creation of a network of regional offices with the head-quarters in Leipzig. This brought him into conflict with Karl C. von Loesch, who believed that the *Stiftung* needed to be centered in Berlin and had to become more politically oriented. Loesch pointed out that the conferences could easily turn into "scientific coffee parties" (*wissenschaftliche Kränzchen*) if more politicians were not included.[161]

More frictions between Loesch and Volz developed because Volz felt that Loesch was trying to curtail the influence of the *Stiftung* by publishing the book *Volk unter Völkern* (Loesch and Ziegfeld 1925) and by founding the *Arbeitsausschuß für deutsche Kulturforscher*. The situation was defused because Loesch agreed that Volz could make changes in the organization of the second volume of the book and by a division of labor in a signed agreement between the *Arbeitsausschuß* and the *Stiftung* which stipulated that the *Arbeitsausschuß* was limited to ethnographic work and subordinated to the *Stiftung*.[162]

Volz was determined to expand his power and attempted to use the *Stiftung* to make Leipzig the center for the production of nationality maps. On 29 July 1926, he contacted the *Statistisches Reichsamt* (Census Office). He pointed out that the common practice of using areal coloring for nationality maps, especially in atlases, had been disadvantageous to the German cause. As a solution he presented three maps, which he said were products of the *Mittelstelle*, and argued that the census of 1925 presented the unique opportunity to depict the nationality distribution on a new statistical basis.[163] Since incontestable cartographic representation methods were required, he offered that the *Mittelstelle* – in collaboration with the Census Office – would prepare nationality maps which would be scientifically correct and advantageous. These maps could then be produced at a scale which made them applicable for use in atlases and distributed to map publishers, who would have no choice but to accept these superior products. As a result, the German cause could be argued much more forcefully at home and abroad.[164]

Since the maps listed in Volz's letter were not made by the *Mittelstelle* but under his direction by the Geographical Institute in Leipzig, Volz's suggestion can only be interpreted as an attempt to use the impartial name and reputation of the *Mittelstelle* to secure more work and influence for himself and the Geographical Institute in Leipzig where he was a professor. However, the Census Office refused the offer very determinedly, even though Volz had managed to get the approval of the Interior and Foreign Ministries for his suggestion.[165]

THE DEMISE OF THE "*STIFTUNG*"

Towards the latter half of the 1920s, Volz came into conflict with other members of the *Stiftung*, in particular with Albrecht Penck and Friedrich Metz. Both of these geographers had a *völkisch* orientation and resented Volz's more environmentally based concept of the German nation. The rift became visible in 1927 during a meeting of the *Stiftung*'s administrative council and also during the conference in Regensburg (Fahlbusch 1994: 76, 116). Things took a turn for the worse in 1928 when Metz openly criticized Volz in a memorandum he circulated among members of the administrative council and the associated council of the *Stiftung*. He questioned the integrity of Volz's research, claiming that it was contradictory and harmful to German demands for more territory.[166] Volz terminated Metz's position with the *Stiftung* the next day during a meeting of the administrative council (Fahlbusch 1994: 87). Neither Penck nor Metz attended this meeting. Apparently Penck was traveling in the United States and he was outraged that the decision about Metz was made without consulting him, but he could not reverse it.[167]

Metz continued his efforts to discredit Volz's reputation even after he received a professorship in geography at the University of Innsbruck in 1929. When the *Stiftung* was organizing another conference in Graz, he sent out a circular to his Austrian colleagues which claimed Volz did not consider Austria to be part of the German region (Fahlbusch 1994: 91). The conference still turned out to be a success, but Metz was unrelenting: he polemicized against Volz at the convention of German geographers (*Geographentag*) in Danzig in 1931 and initiated a campaign by the Nazi press against Volz which lasted until 1933 (Fahlbusch 1994: 92–4). In the end Metz's activities bore fruit. Amidst accusations of embezzlement of funds, the Ministry of the Interior decided to dissolve the Stiftung on 8 August 1931.[168]

Metz planned to continue the work of the *Stiftung* in line with his *völkisch* views and attempted to transfer the library and archive of the dissolved *Stiftung* to a new institute in Berlin, but without success.[169] There was no direct successor organization to the *Stiftung*, but some of its activities were taken over by regional research institutes, such as the *Volkswissenschaftlicher Arbeitskreis* and the *Volksdeutsche Forschungsgemeinschaften*;[170] with this regional fragmentation the *Stiftung*'s integrative function as a coordinator of map production and design to maximize German territorial demands was lost and it was not until several years after the Nazi advent to power that a similar coordination was achieved again (see Chapter 8).[171]

6

MAPS AS WEAPONS
The development of suggestive cartography

But the good intention to print maps is not enough. Rather, what is at stake here is . . . to initiate, through the training of particularly qualified persons, the creation of a special "psychological" or "suggestive" map, which is suitable for enlightenment.

(Ziegfeld 1926: 728–9)

In the aftermath of the First World War, Germans became increasingly aware of the persuasive or suggestive power of maps. This new interest focused on the form of representation and therefore complemented the other politicizations of cartography which addressed the factual content of maps. While the concern for accuracy (Chapter 3) and the development of new concepts of national territory (Chapter 4) were directed at the data, suggestive mapping attempted to make the message of maps more convincing and the dissemination more effective. The idea of suggestive cartography was to change maps from "passive, matter-of-fact depictions into dynamic and impassioned weapons" (Ziegfeld 1935: 244; Schumacher 1934: 651).

Thus, suggestive cartography was from the outset explicitly identified as an act of propaganda by its proponents. This stands in marked contrast to the view held by the people behind the other politicizations of the cartography of national territory, who understood their work as objective and scientific. Clearly, all politicizations should be called propaganda because they were all part of a "deliberate and systematic attempt to shape perceptions, manipulate cognitions, and direct behavior" (Jowett and O'Donnell 1986: 16). The difference is that suggestive cartography was conscious of this, while the other politicizations of cartography (greater accuracy and new concepts) tried to deny responsibility for their political activism. These pretensions of objectivity proved to be fortuitous for geographers after the Second World War. They could point to the "scientific" nature of their work and abdicate any political involvement with the Nazi system.[172]

The inspiration for suggestive cartography came from the examination of foreign maps and the German belief in the superiority of Allied propaganda. When German scientists worked on the development of more accurate ethnographic maps, they noticed that foreign maps paid more attention to effective presentation than those of German origin (see Chapter 3). To people associated with the military, this did not come as a surprise. German military leaders, such as Ludendorff, were convinced that Allied propaganda was very effective during the war and used this as an excuse for their military failures (Ludendorff 1920: 285). Germany was defeated, they argued, because the revolutionary events in 1918 had demoralized the nation. This made Germany easy prey to foreign propaganda. As a result, Germany lost faith in herself, gave up, and signed a humiliating peace: the army, which had been victorious on the battlefields, lost because of internal dissent; it had been stabbed in the back (Eyck 1954 I: 187–91, Morris 1982: 45–6).

The notion of the stab in the back was perpetuated by the entire spectrum of the political right in the Weimar Republic – including the Nazis – and became known as the "*Dolchstoßlegende.*" It not only provided an explanation but also entailed a mission: if Germany wanted to reassert its lost status and become a world power again – which was the prime goal of the political right – it had to make its propaganda more effective, particularly if it wanted to wage a war to achieve this goal. As a first step, the population had to be convinced that the peace treaties were unjust and that Germany deserved more territory.

Two groups believed that maps were most suitable for this purpose: followers of *Geopolitik* and *völkisch* nationalists. The school of *Geopolitik*, which was based on the state-organism theories of Friedrich Ratzel and Rudolf Kjellén (see Chapter 4), became popular after the First World War. Its concepts served as a program for Germany's political future: namely the revision of the peace treaties based on "geographical realities."[173] An important function of *Geopolitik* was political education. Geopoliticians were convinced that the outcome of the First World War was partly due to the inability of politicians and of the general public to understand the importance of geographical factors. Karl Haushofer, a retired general and professor of geography at the University of Munich who was instrumental in popularizing *Geopolitik*,[174] thought that "*Geopolitik* was urgently necessary for the 'dumb people' of Germany (*tumbes Volk*), so that it would not be cheated out of the fruits of its prowess and its heavy and bloody sacrifices by politically smarter (*gerissenere*) enemies."[175] Maps were an obvious choice for this political education since they readily illustrated the relationship between geographical and political factors which geopoliticians pursued so eagerly.

Völkisch nationalists also supported the development of the persuasive qualities of maps, but out of a different tradition. They were not concerned with the idea of bringing Germany's political territory into agreement with the geographical environment, but with unifying all Germans. The first *völkisch* institutions appeared in the late nineteenth century: the *Verein für das Deutschtum im Ausland* (VDA) and the *Alldeutscher Verband*.[176] Up to the outbreak of the First World War, the movement was not very important and lacked notable support from the German government or the German public (Jaworski 1978: 371).

This changed as a result of the ideological mobilization for the war, which appealed to the solidarity of all Germans, and the losses of German territory in the Treaty of Versailles. The separation of several million Germans from the Reich brought about a heightened interest in Germans living abroad. It also revealed the shortcomings of the previous cartographic work of *völkisch* groups. Ethnographic maps and atlases from the pre-war period, such as Paul Langhans's *Alldeutscher Atlas* (Gotha: Justus Perthes, 1905), which had been disseminated to propagate an awareness of the Germans outside the political boundaries, supported many of the provisions of the Treaty of Versailles.

Völkisch nationalists realized that new maps were needed to argue for a revision of the treaty and to achieve their ultimate dream: the unification of all Germans in Europe in one state. Toward this end they encouraged two things: first, the development of new concepts of national territory which would allow for further-reaching demands than the reference to language use,[177] and, second, the production and dissemination of maps which showed the rightfulness of these demands and the injustice of the Versailles Treaty most effectively. Karl C. von Loesch's 1921 proposal to the German Foreign Office is an expression of these ideas: it argued for more research on German territorial rights and for more effective propagation of the results, though it did not outline how improved presentation could be achieved (see Chapter 5, "First Initiatives").

FIRST INITIATIVES

The first public demands for an active German involvement in cartographic propaganda and the first theoretical impulses appeared in the early 1920s. Joseph März (1921), who worked for the *Reichswehrministerium* (Ministry of War) and the Prussian Administration (Korinman 1990: 157) and who authored books on *Geopolitik*, opened up the discussion.[178] He pointed out the persuasiveness of Allied maps, which he contrasted with the bland, ineffectual, and overly scientific style of German maps. März's observations were essentially based on the cartographic material included in Campbell Stuart's book on British war-time propaganda (Stuart 1920). He posited that maps were a most effective means of propaganda because they are easily remembered: a few words were always sufficient to get the message across and it didn't even matter if there were some linguistic glitches such as labeling Germany as *"Deutsches Land"* (März 1921: 261).

The core of März's article was a description of effective techniques of cartographic propaganda by the Allies. For example, the use of a pair of maps to show how German hegemony in Europe had disappeared: the two maps contrasted the plan of a German-dominated *Mitteleuropa* stretching from Berlin to Baghdad with the small area under actual German control towards the end of the war. The application of concentric circles to portray the increasing reach of British bombers over German territory also received attention: März believed that this technique not only made people realize that even distant places, such as Hamburg and Berlin, were not safe, but also had the effect of causing internal dissent in Germany by inciting particularist movements. März did not reproduce these maps to illustrate his argument.[179]

According to März, falsification of maps, such as changing the geographical location of key cities, was a common practice of the Allies, but he pointed out that it was not even necessary to lie. A clever presentation was the key and he referred to a Polish ethnographic map which depicted only the distribution of Poles, even if they amounted to as little as 1 per cent of the population. Since the map omitted all other national groups it gave the impression of a far-reaching Polish influence. März did not include illustrations of the foreign maps he discussed. Instead, the article was accompanied by a suggestive map attacking the Versailles Treaty which was neither commented on by März nor directly related to the text (see figure 6.1). With the exception of Upper Silesia and the Saar region, which formed their own category, the map identified all territories ceded by Germany, and most territories of the former Austrian part of Austria-Hungary, as either robbed or violated. Thus, the map even laid claim to the Czech areas of Bohemia and Moravia.

März lamented the lack of geopolitical instinct in Germany and offered a general critique of German cartographic practice, such as its being overly scientific and excessively accurate. Yet, he did not suggest any concrete steps except for one specific point: he called on German publishers to show the pre-First World War boundaries on their maps to ensure that the lost German territories would never fade from the memory of the German people. The suggestion was heeded promptly. On 19 May 1921, only a month after his article had appeared in the journal *Gartenlaube*, the convention of German geographers (*20. Deutscher Geographentag*) passed the following resolution:

The convention of German geographers proclaims that it is a national necessity and duty that the link to Germandom of the areas which were torn from the German Empire in the Treaty of Versailles, including the colonies, remains clearly visible in atlases and maps, and advocates that only those works for which this is the case, be used for instruction in all school grades.

(*Verhandlungen . . .* 1922: 120)

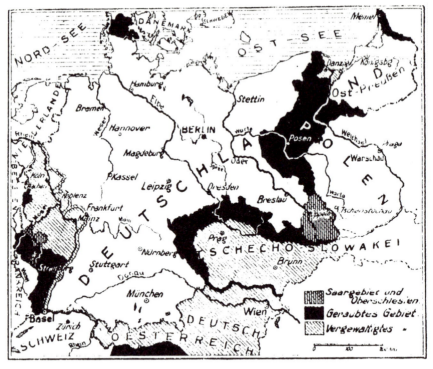

Figure 6.1 Map accompanying März's article, which showed the territorial
provisions of the Treaty of Versailles (März 1921: 261).
The three categories of the legend are (from top to bottom): 1) Saar region and
Upper Silesia; 2) robbed areas; 3) violated areas.

At the convention, Erich von Drygalski, a professor of geography at the
University of Munich, who had initiated the resolution, explained its
importance and necessity. He pointed out that similar demands had been
made before – which might have been an allusion to März's article – and
also mentioned that the Prussian Ministry of Education (*Kultusministerium*)
had issued a decree banning all atlases from use in schools which did not
clearly show the former boundaries of Germany.[180] Drygalski still thought
it was imperative that the German geographers take a stand on this issue
since a number of maps had already appeared which failed to show the
former boundaries. Furthermore, he believed that maps had the utmost
impact on people's perceptions and that it was in the national interest to
show Germany's claim to these areas. Although it was likely that the sale of
German atlases abroad would suffer as a result of this, Drygalski insisted
that it was a worthwhile sacrifice. After all, the claim to these lost territories
had to be upheld to ensure the "victory of truth and of our [Germany's]
right" (*Verhandlungen* . . . 1922: 120). He rejected the possibility of two

different editions of atlases for use at home and abroad since this would give a defeatist message to the Germans living abroad.

März's call for more suggestive maps was picked up again in 1922 by Karl Haushofer. In an article in the journal *Grenzboten*,[181] Haushofer stressed that maps, being inherently geographical, should also become a tool of politics. Referring to the effectiveness of British, French, and Polish maps, he argued that Germans needed to abandon their perfectionism and love of detail and draw maps which concentrated only on the essential issues. This would make German maps powerful tools of education and conviction and would enable Germany to stand up to foreign cartographic propaganda. He pointed out that cartographic lies – which he considered unacceptable because of easy detection – were not necessary: unwanted or unfavorable aspects could be omitted or made insignificant.[182] Selection of facts was the prerogative of the cartographer since he needed to generalize from the abundance of available data.

Haushofer explained that Germans had not made such persuasive maps before because it was not in their national character:

> In contrast to German cartographic representations, the English – because both were a product of the national character, and namely a particularly distinctive one – typified much more and created a more suggestive map image which emphasized the essential and preferably suppressed things that were coincidental or extraordinary; precisely the way England molded its people: individually certainly less attractive and complete, often also less insightful and deep, but more useful for a large and collective purpose: man and map! – life-form on earth and its image.
>
> (cited after the reprint: Haushofer 1928: 343)

Although Haushofer had been more explicit than März about how to make maps suggestive, he fell short of setting an agenda or of pointing out concrete steps needed for its development, unless he intended his statement about national character and maps to mean that Germans should change their character. However, Haushofer was successful in one regard: he popularized suggestive mapping by integrating it into *Geopolitik*.

Haushofer encouraged the use and development of suggestive cartography in the journal he edited, *Zeitschrift für Geopolitik*, which became the main forum for the discussion of general issues relating to suggestive maps in the Weimar Republic.[183] Publications on graphic design, advertising, or psychology generally neglected this media and only mentioned it in passing (Schumacher 1934: 636). As a result, suggestive maps were mostly called "geopolitical" maps. Even though other terms were used – some *völkisch* nationalists favored the name "*volkspolitische Karte*" – the association with *Geopolitik* proved to be more effective: it gave persuasive maps the aura of respectability by making them part of a quasi-scientific discipline. In addition, the use of the term *"geopolitische Karte"* was restricted to representations emphasizing the "psychological effect" (Schumacher 1934: 636), while *"volkspolitische Karte"* designated all maps which addressed political aspects of the German *Volk*, whether they were persuasive or not.

The first truly systematic development of suggestive cartography was vigorously propagated by Arnold Hillen Ziegfeld. Ziegfeld, who was affiliated with the *Deutscher Schutzbund,* was a founding member of the *Deutscher Klub* in Berlin, which was dedicated to the development and propagation of neo-conservative and *völkisch* ideas.[184] He was also an early member of the NSDAP (entry on 25 May 1921!) and while living in Munich (1921–3) he attended meetings in the Hofbräuhaus, Zirkus Krone, and the Sterneckerbräu. Yet, according to a questionnaire by the NSDAP *Reichsleitung, Hauptamt für Erzieher* from 5 May 1936, his *Ortsgruppe* in Berlin-Neutempelhof considered him *streng deutschnational,* a member of the *Stahlhelm,* and noted that he still used the old black-white-red flag in October 1934 and only used the swastika flag after that.[185] His background was mainly in commercial art, but he also was knowledgeable in publishing – he owned the Edwin Runge Verlag – and on questions of Germany's eastern heritage.[186]

In 1927, Ziegfeld, then cartographer and head of the publishing house of the *Deutscher Schutzbund* in Berlin, first attempted to lay the theoretical foundations for suggestive mapping in an article which appeared in the May/June 1927 issue of the periodical *Volk und Reich* (Ziegfeld 1935: 244). He argued that suggestive mapping be dissociated from established cartography and proposed the creation of an entirely new discipline, *"Kartographik."* The term indicated the close relations of suggestive mapping and *"Gebrauchsgraphik"* (commercial art). In a later elaboration on the topic he pointed out that just as the "applied" nature of commercial art was in conflict with "independent" fine art, "applied" suggestive mapping was different from cartography (Ziegfeld 1935: 244). Ziegfeld believed that suggestive mapping and scientific cartography followed entirely different laws and rules:

> Just as it [cartography] as a science and technique is committed to the graphic recording and representation of the surface of the earth and thereby is forced to develop the continuous refinement of its means to the benefit of a more and more faithful naturalism and truthfulness, the suggestive map shall have its function in creating the abstract expression of a slogan, in which the acuteness of the phrased idea and the clarity of the corresponding image combine to the inescapable psychological effect which gives the suggestive map its importance as a political weapon and educational instrument.
>
> (Ziegfeld 1935: 244)

For Ziegfeld it was absolutely essential that suggestive mapping be allowed to develop on its own without restrictions imposed by scientific cartographic practice. He believed that the drawing of good suggestive maps was a matter of inherent talent (*im Blute*) (Ziegfeld 1935: 245). Initially, Ziegfeld took it upon himself to popularize and develop suggestive cartography. He produced numerous suggestive maps in his own unique style and established specialized suggestive mapping offices for the *Deutscher Schutzbund* and *Volk und Reich*. After the National Socialists came to power in 1933, he attempted to give suggestive mapping even more prominence by integrating it into the political education machinery of the Nazi state; however, without much success (see Chapter 9, "Organization of Production").

While Ziegfeld's political activism provided the motivation for suggestive maps – which he started drawing in 1923 – early work with Leo Frobenius, a world-renowned anthropologist, appears to have stimulated his interest in innovative cartographic design. After Ziegfeld's studies in commercial art in Weimar, he had obtained a position at Frobenius's *Institut für Kulturmorphologie* in Munich to work on the *Atlas Africanus* project.[187] Although Ziegfeld was not in charge of the map production for the atlas – he only contributed three articles[188] – he obviously had to deal with the cartographic ambitions which Frobenius had laid out for the atlas.

Citing Friedrich Ratzel – under whom Frobenius had studied – Frobenius (1921: 7) posited that maps were a symbol of reality and therefore, just like pictures, a precursor of language. In addition, maps were static and only showed the situation at a given point in time. This posed problems for subjects dealing with inherently dynamic phenomena, such as culture, which addressed movement and processes. Although map series were helpful in portraying dynamic aspects, processes required more than simply isolating their constituting interrelated elements and then representing them in individual maps. Processes had to be depicted synthetically, that is, their essence could only be "experienced" cartographically if the constituting elements were combined without revealing the interrelation between the elements (Frobenius 1921: 9). However, the atlas fell short of living up to these intentions; none of the synthetic maps were published.

The influence of Frobenius's cartographic ambitions is quite apparent: Ziegfeld stressed dynamic aspects and movement as important characteristics of suggestive mapping (Ziegfeld 1935). Frobenius also seems to have inspired others: the German geographer and geopolitician Otto Maull used the idea of a synthetic representation to distinguish suggestive maps from academic political geographical maps (Maull 1928).

Although Ziegfeld was the most active in propagating suggestive cartography, as witnessed by his articles, official petitions, and his successful establishment of suggestive cartographic offices at the *Deutscher Schutzbund* and *Volk und Reich*, he was only given credit for the actual maps he drew; his other initiatives were forgotten. Karl Haushofer, who had done little for the development of the field except for giving it prominence by making it part of *Geopolitik*, was credited as being the "founder" of suggestive cartography (Schumacher 1934: 635; Jantzen 1942: 354).

THE PRACTICE OF SUGGESTIVE CARTOGRAPHY

Humble beginnings

General propaganda activities by the German government and the military were rather uncoordinated until the very end of the First World War. A central institution was only founded in the spring of 1918, the *Zentrale für Heimatdienst*, which was later called *Reichszentrale für Heimatdienst* with the consolidation of the new republican government (Wippermann 1976: 11–29). Initial propaganda efforts of the *Reichszentrale* focused on the spoken word. This was a continuation of the style of the military information campaigns during the war, which had foregone the use of suggestive images because an appeal to the emotions was considered "unmilitaristic" (Wippermann 1976: 24, 35). As a result, up to the very early 1920s, the use of persuasive maps by government institutions was extremely limited. Just as März had lamented in his 1921 article, official propaganda had neglected persuasive maps, guided by a feeling of "misplaced noble-mindedness and a tiny bit of scientific arrogance" (März 1921: 262).

The government not only failed to take an active role, it even resisted initiatives from private groups. On 8 May 1919, a day after the peace conditions had been presented to the German government, the Gea publishing house in Berlin contacted a member of the German peace delegation at the Foreign Office and pointed out the necessity of producing a map which vividly portrayed the territorial demands of the "enemies." Two weeks later, on 23 May 1919, the publisher wrote to the *Deutsche Geschäftsstelle für die Friedensverhandlungen* (administrative office for the peace negotiations) at the Foreign Office and presented a sample of an actual postcard-size map. The publisher argued that several cities and communities had used depictions of the "outrageous" Allied territorial demands to educate pupils in upper division classes and that the press had demanded government involvement in this regard. To this end, the Gea publishing house suggested the mass production of the postcard-size map which showed the territorial effects of the Treaty of Versailles on Germany. The map was to be based on a poster-size map by the same publisher, entitled *"Die Zerstückelung Deutschlands"* (the dismemberment of Germany), which depicted the areas which Germany had to cede without plebiscite, the areas where plebiscites were to be held, the Saar region, and the areas which continued to be occupied by foreign armies. In addition, some of the economic losses associated with the territorial changes were indicated by bold lettering: for example, "coal" for Upper Silesia, "bread" and "potatoes" for Poznan, and "milk" and "butter" for East Prussia.[189]

The response of the Foreign Office was symptomatic of its passivity in regard to maps of national territory. Just as it had been reluctant to include experts on maps in the peace delegation and to embrace suggestions for mapping projects by geographers (see Chapters 2 and 3), it rejected the proposal, believing that it was too late for the dissemination of the map to have any effect (AA R23076).

Not only were suggestive maps strikingly absent in most government propaganda efforts – the main journal published by the *Reichszentrale, Der Heimatdienst*, contained only a very small number of maps and illustrations up to 1924 – they were also surprisingly ineffective. A map which was included in a publication by the *Reichszentrale* on the Treaty of Versailles (*Reichszentrale* 1919) reveals the carelessness in cartographic design.[190] As can be seen in figure 6.2, it depicted the plebiscite areas and those areas which were to be ceded without plebiscites with two types of shadings, as

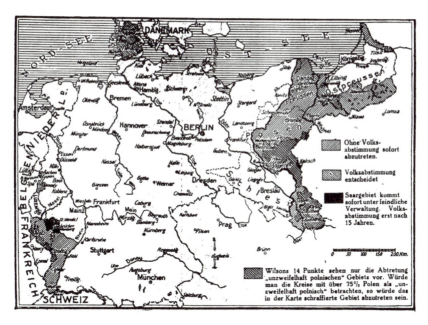

Figure 6.2 Territorial cessions and plebiscites (Reichszentrale 1919).
The four categories of the legend are (from top to bottom): 1) to be ceded immediately without a plebiscite; 2) plebiscite will decide; 3) effective immediately, the Saar region is being administered by the enemy, plebiscite only after 15 years; 4) Wilson's 14 Points only stipulate the cession of "indisputably Polish" areas. If one considers districts with more than 75% Poles to be "indisputably Polish", then the cross-hatched areas on the map would have to be ceded.

well as the Saar region which was to be administered by France for fifteen years in solid black. An additional category of the legend used cross-hatching to designate areas which were more than 75 per cent Polish and argued that only these areas were "indisputably Polish" in accordance with Wilson's Fourteen Points and therefore subject to be ceded. This cross-hatched category not only supported the cession of almost the entire part of Upper Silesia which was given to Poland after the plebiscites, but was also used to delimit areas in Schleswig and Alsace-Lorraine even though it was explicitly stated in the legend that this category outlined 75 per cent Polish majority areas. This shows inadequacies in design and at the same time underlines the fact that the main emphasis of German objections were directed at the eastern provisions of the Versailles Treaty.

Bad design is also apparent in a map produced by the *Reichszentrale* entitled "*Die deutsche Insel Ostpreußen*," which was disseminated by the Press Office (*Presseabteilung*) of the Federal Government to the missions

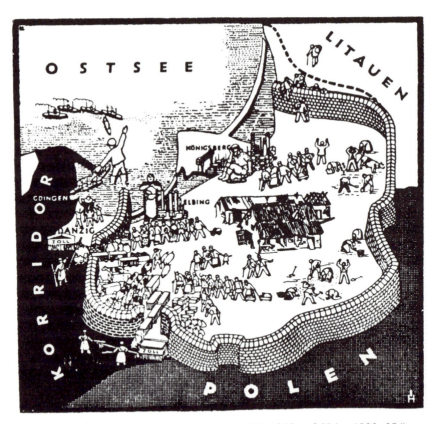

Figure 6.3 The isolation of East Prussia I (Ziegfeld and Kries 1933: 354).

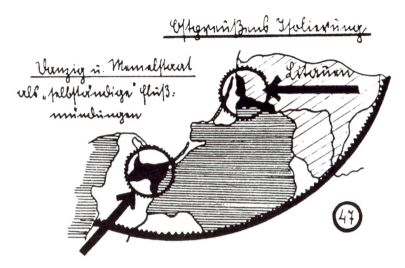

Figure 6.4 The isolation of East Prussia II (Springenschmid 1934: 11).

and consulates abroad. It simply represented East Prussia and the eastern part of Germany in one color and all neighboring territories (which appeared as a uniform area because political boundaries were absent) in another color. Except for the title printed above the map, "The German Island East Prussia," the design gave little support to the message of East Prussia's isolation.[191] Maps in geopolitical and *völkisch* publications were much more effective: they focused on the immediate vicinity of East Prussia and only left a negligible remnant of the German territory with which the province could connect. They also used pictorial elements, such as a wall drawn around East Prussia (see figure 6.3) or half-circles and arrows enclosing the province (see figure 6.4), which gave the impression of a blockade.[192]

Wippermann (1976: 105) points out that the political posters which the *Reichszentrale* produced for its own use or for other official institutions were rather ineffective when compared to posters produced by the National Socialists and other political activist groups. He attributes this to the lack of consultation by experts in political psychology. In questions of design, the *Reichszentrale* worked together with the *Reichskunstwart* Edwin Redslob, which ensured artistic popularity but not a propagandistic message.[193]

A secret memorandum, which was presented to the German Industry and Commerce Association (*Industrie- und Handelstag*) in 1924, also criticized the design of posters by the *Reichszentrale*. It described two posters which were used after the invasion by French and Belgian troops of the Ruhr area in 1923. One poster depicted a French soldier standing menacingly over the Ruhr area, the other showed Marianne, the French national personification, as an Amazon warrior kneeling over the industrial area. Since it was barely noticeable that Marianne had hurt her hand grabbing at the factories, the author argued that the posters were useful to show the military might of France and therefore suitable for French rather than German propaganda.[194]

The lack of attention to maps in official propaganda at the end of the First World War was also pointed out by Wilhelm Volz in a review of his fifty years' work as a geographer (W. Volz 1942).[195] After the peace conditions for Germany were made public in May 1919, the German government organized an exhibition in Breslau entitled *"Deutsche Arbeit und Kultur in Oberschlesien"* (German work and culture in Upper Silesia). Yet, the display failed to include effective maps, which prompted Volz to make an unsolicited contribution: he designed six large wall maps which showed the historical growth of German influence in the area.

The local government was impressed and expressed interest in the maps, a surprising reaction considering the general reluctance of German government offices to become involved in the cartographic representation of national territory. When the district president led a group of British journalists through the exhibition, they "piled up" in front of the maps and studied them intensively for about half an hour (W. Volz 1942: 721). As a result, the local government asked Volz to display the maps in a separate booth (ca. 10m × 10m) at the Easter fair in Breslau. Volz entitled the exhibit *"Unser Oberschlesien"* (Our Upper Silesia). According to Volz, it was so successful that not only was it continually crowded, but its display was continued after the fair. It lasted until the weather became too cold in the fall and Volz estimated that more than 100,000 people visited the booth in the summer alone (W. Volz 1942: 722).

On the whole, there were only isolated and crude attempts to use maps explicitly for political persuasion in the immediate aftermath of the First World War. Most cartographic activity during this period was devoted to the development and production of more accurate ethnographic maps as documentary evidence for revision and the development of alternative concepts of national territory.

Under the auspices of *völkisch* groups

In the mid-1920s, the situation began to change: more and more suggestive maps appeared and some institutions even established cartographic divisions for this purpose. Among the practitioners of suggestive mapping were several publishing houses,[196] as well as the *Deutscher Schutzbund, Volk und Reich*, the *Verein für das Deutschtum im Ausland*, and the *Reichszentrale*, which disseminated maps through its associated institutions, the *Deutscher Lichtbilddienst* and the *Zentralverlag*. Their activities were not isolated, but the product of a personal and organizational network. Some of these connections were an outgrowth of the propaganda for the plebiscites and the response to the French and Belgian invasion of the Ruhr in January 1923. Because official governmental intervention in the plebiscites was prohibited in the Versailles Treaty and in order to camouflage government involvement in propaganda in general, the *Reichszentrale* cooperated closely with private groups, in particular with the *Deutscher Schutzbund*, which had been founded in 1919 to coordinate the work of regional *völkisch* associations.[197] In addition, the *Reichszentrale* shared its building and telephone system with *völkisch* organizations, such as the *Deutscher Ostbund* and the *Reichsverband heimattreuer Ost- und Westpreußen* (Wippermann 1976: 314, fn. 16).

Even though the *Reichszentrale* was somewhat involved in suggestive mapping, it was mainly private initiatives which accounted for the development of this cartographic genre and the production of the vast majority of maps. The core of the connections centered around *völkisch* organizations and private interest groups. The Weimar government supported some of these groups financially, but it did not give directions for their activities. In addition, the groups were not dependent on government funding since some also received funds from German industry, such as the *Wirtschaftspolitische Gesellschaft*, Berlin,[198] and the *Stiftung Volk und Reich*.[199]

The institutional network behind the development of suggestive mapping included in particular the *Deutscher Schutzbund*, the VDA, *Volk und Reich*, and the *Wirtschaftspolitische Gesellschaft*. For example, the VDA took part in the founding of the *Deutscher Schutzbund* and the journal *Volk und Reich* appeared for several years in the publishing house of the *Deutscher Schutzbund*.[200] The *Wirtschaftspolitische Gesellschaft* collaborated with *Volk und Reich* on the production of a small atlas on the East which was to be distributed in different languages and with the *Arbeitsgemeinschaft für Grenzlandarbeit in Westpreußen* on the production of comparative map postcards depicting corridors through other countries, such as an Irish corridor separating Scotland and England.[201]

A key figure in the establishment of such links was Arnold Hillen Ziegfeld. Ziegfeld moved to Berlin in the summer of 1923 to work at the publishing house of Kurt Vowinckel, a professed National Socialist. The Vowinckel Verlag, which paid particular attention to suggestive maps in its publications (Schumacher 1935: 248), was also the main publisher of works from the newly emerging field of *Geopolitik*. The *Vowinckel Verlag* began issuing the journal *Zeitschrift für Geopolitik* in 1924, which means that Ziegfeld as the publishing director must have been involved in the initial negotiations with the chief editor of the journal, Karl Haushofer, who was an influential person in nationalist circles in Germany.

When business went bad at the *Vowinckel Verlag* in the summer of 1924, Vowinckel arranged for Ziegfeld to meet with Karl C. von Loesch, the head of the *Deutscher Schutzbund*.[202] After organizing propaganda efforts for the post-Versailles plebiscites, the *Schutzbund* continued to act as a parent organization for most *grenz- und auslandsdeutsch* organizations, which sought to integrate Germans living abroad with the Germans at home. Ziegfeld started work at the *Schutzbund* on 1 August 1924, devoting most of his time to the publication of the *Taschenbuch für das Grenz- und Auslandsdeutschtum*, which contained numerous suggestive maps. He soon advanced to become director of the publishing house of the *Schutzbund* and established an office for the production of suggestive maps.[203] In addition, he worked in the main office of the *Schutzbund*, where he was in charge of maintaining contacts with individuals and organizations involved in "scientific and literary work on Germandom" (*Deutschtumsarbeit*).[204] His exposure to and interest in issues of Germandom was not just a result of his work at the *Schutzbund*, but also a matter of personal experience. Ziegfeld was born in Japan, where he lived until the age of 10, and both his father and brother spent many years abroad and were active in propagating their German identity.[205]

Economic constraints on the publication activity of the *Schutzbund* as well as personal differences led Ziegfeld to leave his position there in the summer of 1928.[206] He continued to publish works with suggestive maps through his own publishing house, *Edwin Runge Verlag*, and became a freelance for the journal *Volk und Reich*, which appeared in cooperation with the *Schutzbund*.[207] By January 1930, he had advanced to a full-time position in charge of cartographic production at *Volk und Reich*. In addition to heading the suggestive mapping office which he had established at *Volk und Reich*, Ziegfeld was also a member of the editorial board of the journal and involved in other publications of the firm. His efforts were a success: *Volk und Reich* became famous for its suggestive maps (Schumacher 1935: 248, Jantzen 1938: 324).

Ziegfeld not only established offices for suggestive cartography at two different publishing houses, but he made a name for himself as a master of suggestive mapping: his maps were used extensively by a variety of institutions, such as the VDA, *Vowinckel Verlag*, *Verlag Grenze und Ausland*, and also by the *Wirtschaftspolitische Gesellschaft*. Karl Haushofer was so impressed with Ziegfeld's cartographic work that he attested Ziegfeld's method to be "the most promising [approach] . . . in suggestive cartography to achieve the ultimate in the combination of scientific accuracy with practical usefulness for a broad appeal."[208]

Ziegfeld's attempts to increase the production and dissemination of suggestive maps were aided by his association with influential activists in the *Deutscher Schutzbund*, *Deutscher Klub*, and *Volk und Reich*, such as Max Hildebert Boehm, Karl von Loesch, and Hermann Ullmann.[209] His collaboration on book projects with Friedrich Heiss (Heiss and Ziegfeld 1933) and Karl von Loesch (Loesch and Ziegfeld 1925) brought him into contact with many geographers, such as Albrecht Haushofer,[210] Walter Geisler, Albrecht Penck, Waldemar Wucher, Robert Sieger, Fritz Jäger, as well as historians and other researchers working on the East, such as Gustav Aubin, Erich Maschke, Manfred Laubert, and Martin Spahn.

Success

By the late 1920s to early 1930s the private initiatives to promote suggestive cartography showed striking results: there was a "virtual flood" of suggestive maps, which were used in "every public lecture, every newspaper, and in countless books" (Schumacher 1934: 635–6).[211] Even atlases appeared which were solely devoted to this cartographic genre, such as Braun and Ziegfeld (1927) and Schmidt and Haack (1929).[212] Key features of these maps were simplistic design, a bare minimum of information – for example, place names were reduced to the number necessary for orientation – and a lack of supporting evidence for the claims made. They were usually drawn in black and white, or a few select colors (particularly red), omitted details, such as scale bars and legends, and chose dynamic symbols (e.g. arrows) and bold shading. This gave them great advantages over more scientific depictions: they could be produced very inexpensively and easily reproduced in varying sizes.

Realizing these advantages of suggestive maps, the VDA decided to abandon the plan for a regular "scientific" atlas of the distribution of Germans in border regions and abroad (*Grenz- und Auslandsdeutschtum*). It feared that the high costs of a scientific atlas could not be recouped through sales and decided instead to prepare a collection of suggestive black-and-white maps by Friedrich Lange. Lange, an associate judge of a regional court (*Landgerichtsrat*), was so proud of his double doctorate title that he always included it when he signed his maps. His most commonly used sign was "Dr. Dr. F. Lg." The project was started in 1928 and the collection was subsequently published as the small atlas *Volksdeutsche Kartenskizzen*. In 1937 this atlas was in its fourth edition. The VDA used the individual maps before the atlas was completed: they were reprinted in various VDA journals and passed on to newspapers in copy-ready form (for an example, see figure 6.5).[213]

Ease of reproduction also facilitated mass dissemination and use in education. About 100,000 copies of the comparative corridor postcards of the *Arbeitsgemeinschaft für Grenzlandarbeit in Westpreußen* had been disseminated from the time after the end of the First World War until 1932.[214] And starting in the late 1920s, schoolteachers could rely on an increasing number of inexpensive suggestive atlases for incorporation into their curriculum.[215] These atlases were intended for use in education "to show the present-day position of the German fatherland" and "to make German youth understand . . . the future tasks of . . . [the] nation (*Volk*)" (Schmidt and Haack 1929: iii). The simplistic design of the maps made it easy to redraw them on the blackboard or in notebooks and such use was

Berlin als dreifache Grenzstadt

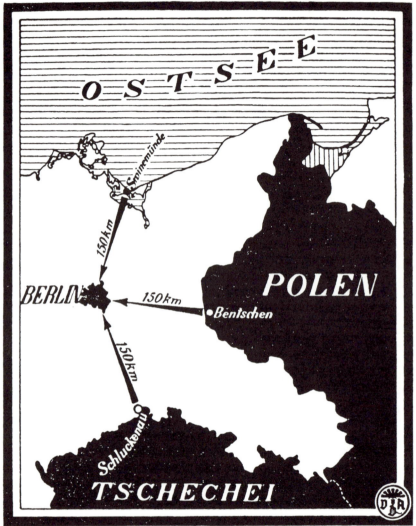

Figure 6.5 Berlin, a border city on three sides (Lange 1937: 27).

encouraged since it improved memorization.[216] Even publications appeared which gave detailed instructions on how to draw suggestive maps for use in education (e.g. Gürtler 1931). Such applications of suggestive maps were given further prominence in 1932, when a proponent of *Geopolitik* included suggestive maps as part of his demand to introduce geopolitical reasoning in primary schools (Thies 1932: 630–3).

The pervasiveness of suggestive cartography was striking. Atlases which had not significantly altered their designs for decades, such as *Putzgers Historischer Schulatlas*, began to use arrows and other suggestive symbols in the early 1930s to make their maps more dynamic and persuasive (Wolf 1978: 708). The new medium was happily shared in the "national interest." Publishers offered slide series of the maps reproduced in their books and even supplied accompanying texts for use in public lectures.[217] The VDA and the *Deutsches Ausland-Institut* in Stuttgart appended a copy-ready map collection (*volksdeutsche Kartenmatern*) to its joint weekly press service. The maps were drawn by Ziegfeld and Lange (Jacobsen 1970: 140). *Volk und Reich* had a similar map service. The 1939 catalog of *Volk und Reich* mentioned that the publishing house had a collection of several thousand maps from its previous publication activities which could be ordered through the "*Volk und Reich Kartendienst*."[218] Starting on 1 January 1930, the Deutscher Lichtbilddienst issued a monthly collection of an average of thirty maps on high-quality art paper at a size of 10.5 × 14.8 cm, ready for reproduction.[219]

Although there was an abundance of suggestive maps in the late 1920s and early 1930s, and many people attempted to draw them, there were only a few cartographers whose maps were widely disseminated because they had pioneered and mastered the craft. Ziegfeld was by far the most prolific. His maps appeared in countless works and he even trained or assisted other cartographers.[220] He was most noted for his bold use of shading (Schumacher 1935: 250), which was emulated by Dora Nadge, who carried out much of the cartographic design at *Volk und Reich*.[221]

The style of the Austrian Karl Springenschmid, who authored several atlases (1934, 1935, 1936), was very distinct because of his preference for handwritten sketches in *Sütterlin*, a uniquely German handwriting which was officially introduced in German schools between 1935 and 1941 and utterly indecipherable without instruction.[222] It is quite surprising that Schumacher credits Springenschmid as the creator of suggestive symbology, since Springenschmid published very few works before 1933.[223] In addition, there were Kurt Trampler, who pioneered the selective use of color, and Friedrich Lange, who also made large-scale maps, such as the *Sprachenkarte von Mitteleuropa*,[224] which was used in the educational programs of the VDA (Jacobsen 1970: 255).

The theoretical development of suggestive maps lagged far behind its practice; it was generally a trial-and-error approach. There was an attempt in the late 1920s to define suggestive maps and how they were different from other cartographic forms (Maull 1928). But this only reflected the generally fruitless endeavor to define *Geopolitik* and how it differed from other fields, such as political geography: Maull argued that "geopolitical" maps, which he defined as applied and containing a political program, were the "theoretical fruit" of political geographic research (Maull 1928:

339–40). Since there was considerable overlap between geopolitical and political-geographic maps, he ended up using a conglomerate term which obliterated the distinction completely: in his conclusion he called them "political-geographic-geopolitical" maps (*politischgeographische-geopolitische Karten*) (Maull 1928: 342).

When the Nazis came to power in 1933, a well-established network existed for the production of suggestive maps. It centered on *völkisch* organizations and it had succeeded in popularizing the application of maps for political propaganda. Suggestive maps were widely disseminated and used a variety of media. Even though most maps were designed effectively because they were the work of a few experienced and talented suggestive cartographers, uniform rules had not been established. Suggestive cartography was powerful, but still somewhat diverse and mostly a private-sector enterprise with little government involvement.

7

CREATING A UNIFIED MESSAGE

By exercising so-called objectivity, the German is only too inclined to advocate literally the position of the enemy. Yet, in all scientific objectivity, we want to avoid hurting ourselves: not all things have to be printed.

(Wilhelm Volz 1924)[225]

The politicization of the mapping of German national territory in the aftermath of the Versailles Treaty – that is, the concern for accuracy, the development of new concepts, and the attention to suggestive representations – had a profound effect on the production and availability of maps. Not only was there a tremendous increase in the number and types of maps of national territory being published, but there were also changing patterns in the messages they carried.

CHANGES IN MAP DESIGN

As even the most nationalistic German scholars had to concede, many German ethnographic maps, even in the 1920s, showed a distribution, particularly in the eastern parts of Germany, which supported a cession of lands to Poland and other countries on the basis of national self-determination: "Unfortunately, German cartography assisted the harbingers of these activities. More than once, Briand supported his false claims with German maps when he attempted to present the Upper Silesian situation in a light which was detrimental to Germany" (Stahlberg 1921: 7).

Stahlberg was referring to *Andrees Handatlas*, which the French prime minister Briand had used to support the territorial provisions of the Treaty of Versailles. The 1914 edition and even the 1921 edition of *Andrees Handatlas* contained a map insert labeled "*Sprachenkarte von Posen und Westpreußen*," on the double-page spread map of Pomerania and Poznan (pp. 47–48). It used areal coloring for the distribution of the Polish and German languages in the area and clearly depicted a broad corridor of Polish-speaking districts from Upper Silesia to the Baltic coast. But when the 1930 edition was published, the map insert was omitted and substituted with a large-scale detail of the Pomeranian Bay (Pommersche Bucht).

Similarly, the ethnographic map of Central Europe in the 1914 and 1921 editions of *Andrees Handatlas* (p. 31), which showed Polish majorities south of East Prussia, in the corridor area, in Poznan, and in much of Upper Silesia, was drastically altered in the 1930 edition: the entire area inside the pre-Versailles boundaries which was depicted as Polish (green) in the 1914 and 1921 editions was now represented in hatchings of green, brown, and red, indicating a mixed distribution. Green designated Polish, red German, and brown the newly introduced categories of the Kaschubes and Masures. Since the brown and red hues were similar, the areas occupied by Kaschubes (Baltic coast) and Masures (southern part of East Prussia) appeared to belong to the German category.

A similar design change took place in *Westermanns Weltatlas*. In the 1922 edition, the map *"Völker, Mundarten, Hausformen"* (p. 32) showed large Polish areas in the eastern parts of pre-First World War Germany and designated the Kaschubes with the color for Poles (green), while the 1932 edition partly shaded some previously Polish (green) areas in the German color (red). The areas which were reclaimed in this manner as mixed German/Polish territories were all designated with minority groups: Kaschubes along the Baltic coast, Masures in the south of East Prussia, and *"Wasserpolen"* in Upper Silesia. Thus, by the early 1930s, that is, before the Nazis came to power, German maps had been altered sufficiently not to contradict claims for revisions of the treaties. What is striking about these cartographic Germanizations is that German geographers in the late 1920s warned vehemently about the Slavic threat which the lost German territories in the East were exposed to (e.g. Wittschell 1926, Obst 1929). They argued that Germans living outside Germany were subjugated to brutal slavicization and predicted an eradication of German culture and language in these areas.[226] So it is quite surprising that there were suddenly more Germans depicted on the maps when supposedly there were fewer and fewer of them left.

School atlases

A deeper insight into the design changes in German maps can be obtained by analyzing school atlases. In comparison to other cartographic artifacts, such as wall maps, general atlases, and maps appended to books, school atlases usually not only have a wider dissemination, but are studied carefully in the classroom and have more frequent editions.

The impact of school atlases on the general population is best expressed by the following statement by K. Frenzel, the chief secretary of the *Deutsche Kartographische Gesellschaft* (DKG):[227]

Already the first map, which is placed into the hand of the 8- to 9-year-old child, the school map (*Schulkarte*) is a product of private carto-graphic firms. And how great is the impression or rather influence of the first map on man! . . . But for many people, the school atlas indeed continues to be the only atlas for their entire life.

(Frenzel 1938: 367)

Two of the most influential school atlases in Germany were *Putzgers Historischer Schulatlas* for the subject of history and *Diercke Schulatlas* for the subject of geography. The Putzger atlas, which was first published in 1877, is considered "the" historical school atlas of Germany and Austria and boasts one of the highest print runs in the world (Wolf 1991: 21–2). The Diercke atlas did not have such an overwhelming control of the market – it had competitors, such as the Velhagen & Klasing school atlas – but its print run is still impressive: between 1919 and 1932, a combined total of more than 720,000 copies of the large edition of the Diercke atlas were sold.[228]

The analysis of the yearly editions of the Diercke atlas between 1918 and 1932 allows for a more precise dating of design changes.[229] The ethno-graphic map "*Deutsches Reich, Völker- und Sprachenkarte*" (German Empire: map of peoples and languages, 1:7,500,000) on page 140 of the different Diercke atlas editions used green hues to designate Polish and other Slavic languages and red hues for Germanic languages. The editions of this map for the years 1918 to 1925 still supported the eastern territorial provisions of the Treaty of Versailles: the green hues for the Slavic language areas covered much of Upper Silesia, parts of East Prussia, and reached the Baltic Sea in the corridor region. The map in the 1926 edition clearly showed a different distribution: the areas in the corridor, the southern part of East Prussia, and Upper Silesia designated as Polish (green) in the 1918–25 editions were now portrayed as a mixed area: the green was partly covered with hachures and dots in a reddish-brown hue and labeled as areas of Kaschubes, Masures, and "*Wasserpolen.*" This distribution remained unchanged up to and including the 1932 edition.[230]

The process of a gradual cartographic Germanization of the lost terri-tories is corroborated by an examination of *Putzgers Historischer Schulatlas*. The editions of the Putzger atlas during the Weimar Republic (42nd edition, 1920, to 50th edition, 1931) featured a map of the distribution of peoples and languages in Europe at the end of the nineteenth century.[231] Since this is an historical map, which depicts the ethnographic make-up of Europe for a given point in time, variations in different editions of the map cannot be attributed to actual changes which might have occurred during the time that elapsed between publications. Thus, the increase in German territory which was most pronounced between the 47th edition (1926) and the 48th edition (1928) could not have resulted from availability of more recent data, but simply represented the attempt to portray a more favorable situation for Germany. The Putzger atlas went even further in reclaiming lost German territory than the Diercke atlas: starting with the 1928 edition, the Putzger atlas designated all lost territory in the East, including the district of Poznan, as mixed German/Slavic.

The changes in these maps reflect the demands of German geographers who had worked on the development of more accurate ethnographic maps in the early 1920s (see Chapter 3). The geographers had rejected the use of areal coloring and demanded that the Kaschubes, Masures, and other minority groups be distinguished from the Poles. However, the revised maps did not adopt the novel forms of representation for ethnographic maps developed by geographers. Instead, following Penck's (1921a: 171) suggestions, cross-bars or interspersed colored dots were employed to indicate mixed ethnic distributions. One of the reasons might have been cost: while a change in the depiction of minority groups could be accomplished fairly easily, it would have been very expensive to completely redraw the maps. In addition to being cheaper, this solution also allowed for a gradual change which was not as noticeable and therefore less suspicious.

Judging from the changes that occurred in all atlas maps which were examined, especially the frequent editions of the Putzger and Diercke atlases, the most marked cartographic Germanization took place around 1926/7 and also was restricted to the eastern boundaries; the depiction of ethnographic distributions along the new boundaries with Denmark, Belgium, and France remained virtually the same.

Ethnographic wall maps

The timing and type of change observed in atlas maps matches well the revised, i.e. more Germanic, representation of two ethnographic wall maps published by the Westermann Verlag. The uniformity of the change in all these different maps suggests some sort of government intervention. However, the correspondence between the publisher and the map author K reveals that the general increase in areas depicted as German was not the result of censorship but rather of pressure exerted by private interest groups.[232]

In early 1921, the Berlin representative of the Westermann Verlag was told by a government building officer (*Regierungsbaurat*) from Küstrin that K's ethnographic map of Germany (at a scale of 1:1,000,000) contained many mistakes and was in some areas completely false. When the publisher requested more specific information, the building officer replied that the map depicted areas as being Slavic which had voted to remain with Germany in the recent plebiscites. He also informed the publisher that he had brought the map to the attention of the *Deutscher Ostbund*, an organization representing refugees from Poznan and dedicated to the preservation of German interests in the eastern territories.[233] The publisher also received a complaint about the map from another interest group, the *Ostmarkenverein*, and informed K about the situation. In his reply to the publisher on 5 July 1921, K defended the scientific integrity of his map. He pointed out that his map showed the distribution of peoples based on language use. Since the plebiscite results in his view had nothing to do with language, he could not use them if he wanted to be consistent and objective in his representation. He added that he would welcome any other constructive criticism and suggestions as well as material supplied by the *Ostmarkenverein*.

In October 1922, the publisher again wrote to K and informed him of complaints by the *Deutscher Schutzbund*,[234] the *Ostdeutscher Heimatdienst*, and others. The publisher also mentioned that he had been in contact with the cartographic office of the Prussian census (*Plankammer des Preußischen Statistischen Landesamtes*),[235] which had also become concerned about language data. The office had published the results of a language census (*Muttersprachenzählung*), which it now recognized as a mistake because the results were being used against Germany. Since the maps by K and other ethnographic maps touched upon similar national issues, the publisher suggested that K get in contact with his professional colleagues to arrive at a consensus and clarification.

On 9 August 1924, the publisher informed K that a bookstore had returned a copy of K's ethnographic map of Western and Central Europe because the director of a school who had ordered the map through the bookstore found it to be false and anti-German. K replied on 10 September 1924, mentioning that he had also written to the director of the school. He reiterated his earlier statement that it was only possible to base a scientific ethnographic map on the distribution of languages and that the use of other criteria, such as the national will or cultural affiliation, would require a complete redrawing of the entire ethnographic map of Europe. He stressed that the language of the Masures was indisputably Polish and that he therefore could not represent them as speaking German. In K's view it was the task of the teachers to point out to the schoolchildren that in the East, Germany had been able to win over non-German-speaking peoples to the German culture. As if to show his nationalist sentiment, he lamented the lack of attention given by the government to Germans abroad, who as a result often did not succeed in preserving their language as the Masures had been able to.

Only four months after K's reply, the publisher furnished him with further complaints in a letter dated 9 January 1925. The publisher had exhibited K's two ethnographic maps at a convention of the Prussian teachers' association in Kassel, which was attended by more than 400 school directors and teachers from districts all over Germany. The maps were criticized repeatedly and despite the publisher's explanation that the maps had been prepared on the basis of the most recent official statistical material, which did not allow for a different representation, two "very influential" persons, a school director from Breslau and a teacher from Berlin, could not be appeased and demanded a detailed justification by K. The publisher asked K to comply with their request and also to supply him with a comprehensive exposition, which he intended to give to his sales representatives so they could respond effectively to future criticism.

Once more, in a letter of 24 February 1925, K insisted on the scientific integrity of his map and pointed out that neither the Kaschubes nor the Masures spoke German and that they therefore could not be portrayed as such on his map. He mentioned that he would write to the school director and the teacher to find out the nature of their criticisms and that he was going to search for additional data. In addition, he remarked:

> It is after all a good thing in itself that the interest in the East [is] so great at the present, and it would have been good if it had existed earlier, before the war, when most people had very unclear ideas about the distribution of Germans in the borderlands and in the eastern states. Now, there seems to be the tendency to think that it is much greater than it actually is, not to mention the high rate of return migration.

(WWA, v30/217)

Before K's response reached the Westermann Verlag, the publisher mailed yet another letter in which he quoted objections to K's maps. The Silesian teachers' association had contacted the Westermann Verlag and lamented that the representation of the Upper Silesian language boundary "bordered on contempt for the true situation" and that while such maps might be expected from England and France, their publication by a renowned German publisher was absolutely incomprehensible. The teachers' association stated further that it had not informed the public yet, "since it expected that the maps would be withdrawn from circulation immediately." The publisher requested an immediate response from K.

Although there is no further correspondence by K in the files of the Georg Westermann Werkarchiv, a letter by the publisher to K, dated 2 June 1926, and examinations of different editions of K's ethnographic maps reveal that the mounting pressure from interest groups resulted in changes in the maps. In the letter, the publisher informs K that his ethnographic map of Western and Central Europe would soon be sold out and therefore was in need of a new edition. Referring to the numerous complaints about the map, he reminded K that they had somewhat accommodated the wishes of the critics by adding the boundaries of the German Empire as well as the designations "*Masuren*" and "*Wasserpolen*" to the existing map. He also suggested that K might come up with a different solution and that the boundaries of the distribution of Germans should be examined in detail since, for example, the Hultschin area was supposed to have a predominantly

German population even though the existing statistical data seemed to prove otherwise.

Not surprisingly, the second edition of K's ethnographic map of Western and Central Europe, which appeared some time after 1926, showed a marked change from the first edition: the area of the Masures was now depicted in the German red combined with little and hardly noticeable green dots. The Kaschubes, who had been depicted with the same dark green color as the Poles in the first edition, were now represented in a very light green to distinguish them from the Poles. While changes in the distribution of Germans in Upper Silesia were barely visible, the Memel territory, which was previously Lithuanian, was now portrayed as mixed German/Lithuanian. However, the Hultschin area was not represented as predominantly German and the area of the corridor actually showed fewer German-language areas, an indication that K tried to be truthful by portraying the decline of Germans in these ceded territories. The second edition (1928) of K's other ethnographic map – the ethnographic map of Germany – showed similar changes, with the exception of Upper Silesia, in which the entire area within the pre-First World War boundaries of Germany was now depicted as mixed German/Slavic.

A CONCERN FOR BOUNDARIES AND PLACE NAMES

Changes in map design during the Weimar Republic were not just confined to ethnographic distribution, but also included political boundaries and place names. In 1921, calling it "a national necessity and duty," the 20th Convention of German Geographers (*20. Deutscher Geographentag*) had passed a resolution which demanded that only those maps would be allowed for school use which clearly showed Germany's pre-First World War possessions, including her colonies (see Chapter 6, "First Initiatives"). The idea behind it was to keep the memory of the old possessions alive and to stake a German claim for revision. In June 1925, during the 21st Convention of German Geographers in Breslau, the central commission (*Zentralausschuß*) of the convention proudly reported that the resolution had been a complete success and that even commercial map publications followed this practice. There were only a few exceptions, namely maps in newspapers. The commission suggested that geographers use their personal influence to remedy this situation rather than pass a new resolution (*Verhandlungen* . . . 1926: 185).

At the same 1925 convention, German geographers declared it "a national duty" that German place names abroad be preserved and demanded that on all maps German place names be given preference and named first. The resolution addressed all editors and publishers of maps, atlases, travel guides, and all geographical literature. The proponents of the resolution argued that abandoning the German place names threa-

101

tened the continuity of scientific works and would also give the Germans abroad (*Grenz- und Auslandsdeutsche*) "the bitter feeling" that the homeland (*Heimat*) was turning its back on them (*Verhandlungen* . . . 1926: 190–1).

At first glance, the 1925 resolution does not appear very noteworthy or important; it was customary then and still is at times today to name as many localities as possible on maps and in atlases in the language of the country publishing the atlas. In fact, since the resolution even endorsed the use of a second, foreign place name, it seemed to be not only conciliatory, but even superfluous. However, while the practice of naming localities in one's own language is often intended for reasons of better familiarity and to facilitate pronunciation, it is also a political statement. In disputed areas, place names are not just mere identifications of location, but are wrought with political meaning; they constitute territorial claims (Hall 1981: 320–2; Boehm 1932: 83–5). This was particularly true for Eastern and Southeastern Europe, where the intermixed distribution of nationalities produced conflicting place names for many localities (e.g. German: Hermannstadt; Romanian: Sibiu; Hungarian: Nagyszeben) and choosing one over the other was a political statement, a territorial claim (Jaworski 1980: 251). There are recent examples of this: the Polish National Tourist Office Travel Planning Guide 1989/90, *Welcome to Poland*, still included "maps with Polish localities printed in German" in its list of forbidden items, along with narcotics, weapons, pornography, and printers.

Another incident in 1925 underlines the importance of place names for German nationalist aspirations: the publication of Albrecht Penck's concept of the *Volks- und Kulturboden*. In outlining the extent of the *Kulturboden*, place names provided the proof for German historical possessions, that is, for the "dominance of German culture." Therefore, the resolution of the German geographers has to be seen as an attempt to enforce the implicit cartographic message in all German maps that there were still considerable territories abroad which Germans could rightfully claim.

As early as 1916, the VDA had pursued the same goal by demanding that all map publishers in Germany rely on German place names as much as possible – even suggesting the Germanization of English place names overseas. But the VDA had not been successful in evoking a concerted response among geographers or from official institutions.[256] It took the heightened awareness of the political power of maps as a result of the Paris treaties, as well as a nationalist campaign against the publication of a German atlas, for geographers to publicly embrace the issue of German place names in 1925.

The nationalist campaign which ultimately led to the place-name resolution goes back to 1923, when the Ullstein Verlag released a new world atlas, the *Ullstein Weltatlas*. Intended as a reference work for commerce and travel, it was the first comprehensive atlas in Germany which showed the new political situation after the First World War (Jaworski 1980: 251). Given the nationalist feelings associated with place names in Eastern and Southeastern Europe, the enterprise of portraying the changed political map of Europe proved to be a difficult one for the publisher.

The publisher opted to give preference to the place name in the language of the country which was represented and to add German place names where it was important (Jaworski 1980: 252). This was sure to appease the newly formed states, such as Czechoslovakia, which were very sensitive about this issue. In addition, it helped the orientation of the map reader since Czechoslovakia, especially, refused to recognize the old German names in commerce and postal transport (Jaworski 1980: 252–4). Initially, the atlas was well received in Germany and abroad, but nine months after the publication of the atlas, an article in a German newspaper in Czechoslovakia (*Reichenberger Zeitung*) expressed outrage that the German place name of the city where the paper was published had been degraded to second place and even abbreviated on the atlas map of Czechoslovakia.[237] The article also lamented that in several instances German place names had been omitted entirely and that this practice was helping Czech nationalist aspirations.

The complaint was reprinted in several other Sudeten-German papers. It reached an even wider audience when it was published in the *Deutsche Allgemeine Zeitung*, expanded by a scathing attack on the Jewish publisher which implied that he was anti-German and collaborating with France and Czechoslovakia (Jaworski 1980: 253). The attack had been instigated by Hermann Ullmann, who was the editor of the VDA paper *Deutsche Arbeit*, where the expanded article had appeared originally, before it was reprinted in the *Deutsche Allgemeine Zeitung*. The publisher, Franz Ullstein, who was a member of the VDA himself, felt insulted and sued Hermann Ullmann for libel (Jaworski 1980: 253).

Hermann Ullmann welcomed the publicity of the whole affair and exploited its potential for the German nationalist and revisionist cause. As a board member of the VDA and the *Deutscher Schutzbund*, he was well connected in nationalist circles and able to enlist various organizations in the preparation of the court case. Apart from Sudeten-German groups and conservative newspapers, such as the *Tägliche Rundschau* in Berlin, he was able to get support from well-known geography professors, such as Walter Vogel (Berlin), Wilhelm Volz (Leipzig), and Fritz Machatschek (Zurich), who prepared critiques on the atlas map.

Franz Ullstein still won the libel suit, but the nationalist front which Hermann Ullmann had formed continued its activities: the German geography convention passed the 1925 resolution on German place names only one week after the court verdict and the right-wing *Deutschnationale Volkspartei* (DNVP)[238] made a motion in the German parliament (*Reichstag*) to support the resolution of the *Geographentag*, pointing out that the Ullstein atlas incident should serve as a warning (Jaworski 1980: 258). The case did not go to the appeals court since Franz Ullstein and Hermann Ullmann reached a settlement out of court through the mediation of Karl C. von Loesch and others (Jaworski 1980: 258–9).

To appease his nationalist critics, Franz Ullstein published a new map of Czechoslovakia which was given free of charge to the owners of the atlas. The map was at a scale of 1:1,500,000, used German place names, and only depicted the Czech place names on a transparent cover sheet. As in the previously discussed cases, pressure by interest groups – a form of private censorship – had resulted in a change in cartographic design. To ensure that the "correct" place names were chosen in the future, reference works were published by a commission of the *Geographentag* (Gradmann 1929) and also by the *Deutsche Akademie* (Kredel and Thierfelder 1931).

THE "*STIFTUNG*" AS AN ENFORCER OF UNIFIED DESIGN

The *Stiftung für deutsche Volks- und Kulturbodenforschung*, which considered itself a mediator between science and politics, also acted as an enforcer of unified cartographic design. In addition to being involved in the elimination of undesirable maps and their replacement by more favorable versions, such as Muke's map of the distribution of the Sorbs (see Chapter 5, "Scientific-Political Congresses"), it used its connections to the government to bring about change: a map in the 5th edition of *Meyers Konversationslexikon* was publicly criticized in a newspaper article in the *Ostdeutsche Morgenpost* on 17 October 1926, for not portraying enough German territory in Upper Silesia.[239] A regional historian, Freiherr von Richthofen, had discovered the shortcomings of the map and asked Wilhelm Volz for assistance. Volz's reply, which was mentioned in the newspaper article, stated that he had received similar complaints from several private and official institutions and that he had contacted the publisher, the *Bibliographisches Institut*, Leipzig. Volz was certain that the "absolutely false map" would be replaced in the next edition and he expressed his regrets about the whole incident since the *Bibliographisches Institut* was generally completely reliable in such matters.

However, Volz did not just contact the publisher as was mentioned in the article, he made sure that future map publications by the *Bibliographisches Institut* would be controlled by the *Stiftung*. In a letter from the *Stiftung* to the Foreign Office, dated 28 October 1926, Volz stated that he would gladly follow the suggestion of the Foreign Office member P. to become involved in future map publications of the *Bibliographisches Institut*. While this appears to indicate a censorship initiative by the government, i.e. the Foreign Office, the remainder of the letter reveals that the *Stiftung* itself had instigated the involvement of the Foreign Office: Volz remarked that the *Stiftung* had lodged complaints with the *Bibliographisches Institut* for years about their language maps and that it was very helpful that the Foreign Office had exerted its influence on the *Bibliographisches Institut*, just as the *Stiftung* had requested from it and other authorities.[240]

In several other cases, the *Stiftung* was a "consultant" to publishers on the use of appropriate terminology and map design. For example, it agreed in late 1926 at the request of the Minerva Verlag, Berlin, to collaborate in the production of maps depicting German territory.[241] In a letter of 1 February 1927, it pointed out mistakes in the book *Ostpreußen* which was published by Verlag Velhagen & Klasing, Leipzig, and offered its assistance in the preparation of a revised edition.[242] And on 5 April 1930, it met with the authors of the geography textbook *Seydlitz Lehrbuch der Erdkunde* (Hirt Verlag, Berlin) to discuss the design of ethnographic maps and the use of the terms *"grenzdeutsch"* (Germans in border areas) and *"Korridor"* (corridor). Agreement could not be reached on the use of *"grenzdeutsch"* and it was proposed that the *Stiftung* contact the *Schulgeographenverband* (Association of German Geography Teachers) for further discussions. Regarding the term *"Korridor,"* it was concluded that *"polnisch"* (Polish) should not be mentioned together with *"Korridor"* since this could be interpreted to signify the ethnographic make-up of the area. In future, the word *"Weichselkorridor"* (Vistula corridor) was to be used and always in quotation marks. It was also decided at the meeting that any ethnographic maps be given to the *Stiftung* for examination.[243]

The changes in ethnographic distribution, the mandatory depiction of pre-First World War boundaries, the Franz Ullstein affair, and the consulting activities of the *Stiftung*, all underline how in the Weimar Republic of the mid-1920s geographers and *völkisch* groups joined forces to ensure a unified revisionist message in maps. It was a success for both parties: *völkisch* groups achieved greater respectability in their political work through the backing of "independent, objective" scientists, while the geographers benefited by having the national importance of their work publicized. Since many of the attacks on "anti-German" maps were printed in newspapers or otherwise made public, the coercive activities of *völkisch* groups and geographers also had the effect of increasing public awareness of the political nature of maps.

THE EMERGENCE OF NEW MAPS

The elevation of national self-determination to the central principle of the peace treaties sparked a marked increase in the production of maps of German national territory. As the German geographer Carl Uhlig pointed out in August 1919: "Almost every day brings us new maps, which are intended to show the territorial side of the peace conditions – which are so shocking to us – and how it defies the distribution of Germans."[244] The statement reveals the political agenda of new maps: they intended to prove that the peace treaties violated the principle of national self-determination by showing the maximum extent of German national territory.

Composite ethnographic maps

In the years immediately following the peace treaties, the political agenda of exposing how the peace treaties had violated German national self-determination posed a problem. Language-based ethnographic maps were not able to claim the ceded territories in the East, which were ethnically intermixed, even if they employed the most advantageous methods. And maps employing the newly developed concept of geo-organic unity only supported demands for Germany's pre-war state territory, not for ethnic German areas beyond it, such as Austria. As a result, a new type of map emerged, which took inspiration from März's (1921) and Haushofer's (1922) call for more politically motivated maps and combined the advantages of both ethnographic maps and geo-organic maps. This composite map was based on the distribution of the German language, but circumvented the problem of the ethnic mix in the ceded territories by simply choosing a separate category which identified the ceded territories as an integral part of German national territory. The category was labeled "*Deutsches Grenzland in Not*" (German borderlands in jeopardy), "*entrissene Gebiete*" (severed areas), or "*geraubte Gebiete*" (robbed areas), "*vergewaltigte Gebiete*" (violated areas), etc. These maps were inconsistent in the application of their underlying principle – which was supposedly the depiction of the ethnographic make-up of an area – because the intermixed categories had no common denominator. But their purpose was not to give evidence for the German character of these areas, but to portray their exclusion from the German state as an injustice. These maps did not argue, they accused.

Two of the most widely disseminated composite ethnographic maps were *Deutsches Grenzland in Not* and *Das ganze Deutschland soll es sein*. The map *Deutsches Grenzland in Not*, at a scale of about 1:1,500,000 (ca. 80 x 95 cm), was first published in 1920 in Austria and later disseminated in a slightly modified version by the VDA until 1936, when this practice was stopped voluntarily by the VDA as a result of Polish criticism (see figure 7.1).[245] The map depicted German-language islands in Poland, Czechoslovakia, and Hungary, but omitted all non-German areas within the pre-First World War boundaries of Germany. Neither the Polish areas in the corridor region and Upper Silesia, nor the Czech areas in the Hultschin district, the Lithuanian area in the Memel territory, the French areas in Alsace-Lorraine, the Belgian areas in Eupen and Malmedy, or the Danish areas in Schleswig were shown. Instead, the map subsumed German Austria under the same category as Germany and laid claim to the Sudetenland and areas bordering German Austria which it designated as German, such as South Tyrol.

A GEOGRAPHY LESSON FOR YOUNG GERMANY

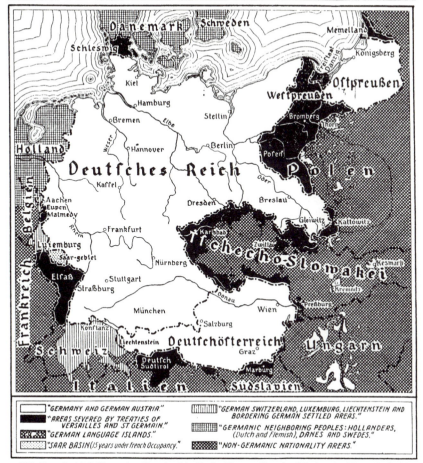

"GERMANY AND GERMAN AUSTRIA."	"GERMAN SWITZERLAND, LUXEMBURG, LIECHTENSTEIN AND BORDERING GERMAN SETTLED AREAS."
"AREAS SEVERED BY TREATIES OF VERSAILLES AND ST GERMAIN."	"GERMANIC NEIGHBORING PEOPLES: HOLLANDERS, (Dutch and Flemish), DANES AND SWEDES."
"GERMAN LANGUAGE ISLANDS."	"NON-GERMANIC NATIONALITY AREAS."
"SAAR BASIN (15 years under french Occupancy."	

This Map, Entitled "Deutsches Grenzland In Not" (German Frontier Districts in Distress), Is Distributed by the Nazi Ministry of Propaganda and Hangs in the School Rooms of Germany.

Figure 7.1 Composite ethnographic map I: *"Deutsches Grenzland in Not"* (*New York Times*, 17 March 1935). (© 1935 The New York Times Co. reprinted by permission.)

Luxembourg, Liechtenstein, and most of Switzerland were also shown as German, albeit with a slightly different color design than the rest. There was also a category for "Germanic states", which included Holland, the Flemish part of Belgium, Denmark, and Sweden. All other countries were subsumed under one heading: "non-Germanic ethnic areas" (*nicht-germanische Volksgebiete*).

Friedrich Lange's map, *Das ganze Deutschland soll es sein,* (ca. 75 × 45 cm), which was first published in 1923 by the *Deutsch-österreichischer Volksbund,* Berlin, showed a similar distribution of German areas to that of the map

Deutsches Grenzland in Not. The only notable exception was that the classification "Germanic states" was omitted, Holland, the Flemish part of Belgium, Denmark, and Sweden were included in the non-Germanic category (*fremdsprachiges Ausland*). Protests regarding this map by other countries

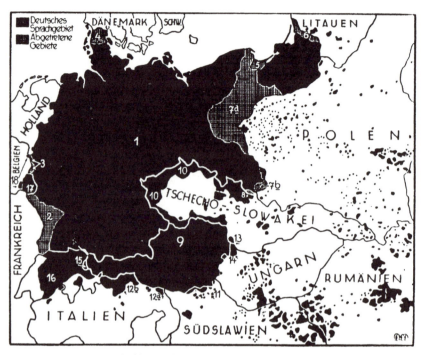

Deutsches Sprachgebiet mit den abgetretenen Gebieten.

1. Deutsches Reich.
2 bis 8: Abgetretene Gebiete.
2. Elſaß-Lothringen.
3. Eupen-Malmedy.
4. Nordſchleswig.
5. freie Stadt Danzig.
6. Memelgebiet.
7a. Poſen-Weſtpreußen.
7b. Oſt-Oberſchleſien.
8. Hultſchiner Ländchen.
9 bis 12: Die Deutſchen des geſchloſſenen Sprachgebie-

tes in den Nachfolgeſtaa-
ten Oeſterreichs.
9. Deutſchöſterreich.
10. Sudetendeutſche.
11. Deutſche in Südſtelermark und
Südkärnten.
12a. Kanal-Tal.
12b. Südtirol.
13 bis 14: Die Deutſchen des geſchloſſenen Sprachgebie-
tes in den Nachfolgeſtaa-
ten Ungarns.

13. Preßburg und Umgebung.
14. Burgenland, öſtlicher Teil.
15. bis 18: Die Deutſchen des
geſchloſſenen Sprachge-
bietes in den anderen
Staaten.
15. Liechtenſtein.
16. Deutſche Schweiz.
17. Großherzogtum Luxemburg.
18. Das deutſche Sprachgebiet in Alt-
Belgien.

Figure 7.2 Composite ethnographic map II: "The German-language area with the ceded territories" (Fittbogen 1928: 12).

The categories of the legend are: "German language area" (top); and "ceded territories" (bottom). The map was reprinted in the 1932 annual report of the *Verein für das Deutschtum im Ausland* (VDA) with the title *Das Deutschtum in Mitteleuropa.* The only difference was that the VDA labeled the category "ceded areas" (*abgetretene Gebiete*) as "lost areas" (*verlorene Gebiete*). The annual report is preserved at the Bundesarchiv (BA R153/92). A similar map also appeared in Fittbogen (1924: 25).

prompted the German Foreign Office to put pressure on the *Deutsch-österreichischer Volksbund* to halt its distribution in 1929.[246]

The pervasiveness of composite ethnographic maps is illustrated by the fact that they were even used by the Weimar government. In 1928, the Weimar Ministry of Information, the *Reichszentrale für Heimatdienst*, published a booklet (*Richtlinie* 70, September 1928) entitled *Die Deutschen außerhalb der Reichsgrenzen*, which was authored by Gottfried Fittbogen, an expert on German national issues who was affiliated with the VDA. As an illustration, it included the map, *Deutsches Sprachgebiet mit den abgetretenen Gebieten* (see figure 7.2). The map was drawn by the *Deutscher Lichtbilddienst*, a picture service associated with the *Reichszentrale*, which offered copy-ready illustrations in a monthly publication.

An emphasis on German cultural achievements

By the mid-1920s, claims to German national territory shifted in emphasis and expanded even further. The catalyst for this development was the publication of the map *Der deutsche Volks- und Kulturboden in Europa* in 1925. The map, at a scale of 1:3,270,000 (ca. 110 × 125 cm), was prepared by Albrecht Penck and Hans Fischer at the suggestion of the VDA.[247] Geared for educational use, it was disseminated as a wall map (RM 25 mounted on cloth, RM 9 regular) as well as in a smaller format as an insert for school atlases (RM 0.10).[248] The map not only allowed for further-reaching German demands than composite ethnographic maps, but its backing by a renowned professor of geography and its scientific terminology gave it considerably more credibility.

The map expanded on the two main categories of *Volksboden* and *Kulturboden*, which Penck had introduced in his article on the concept (Penck 1925; see also figure 4.4). In addition to the German *Volksboden* in solid red and the German *Kulturboden* in red honeycomb symbols, it included four new categories: 1) "areas under 100 years of German administration" in dashed red honeycomb symbols (which encompassed the former Austro-Hungarian Empire, Estonia, and Latvia); 2) "areas of the lower German language in Holland and Belgium" in red cross hachures; 3) "areas lost since 1500 in the West" in closely spaced red hachures; and 4) "areas of German language use in commerce in the East" in widely spaced red hachures.

Since all other land areas were represented in beige, the brightly visible red German cultural region encompassing all these categories dominated Europe in an unprecedented way: its western limits stretched from Brussels to Nancy, Besançon and almost as far as Geneva, while the eastern boundaries followed the line from Lake Ladoga to Smolensk, Charkow, along the Dnjepr to Cherson. The depiction of the post-First World War boundaries in green served as a reminder of how the Paris peace treaties fell short of giving Germany "what it deserved."

The overall impression of the map was considered a success for the German cause:

> The success of such a representation is the impression of the exuberant energy of our people, which being forced to a halt in the West by nature and foreign might, now spread all the physical and intellectual assets it possessed eastward.

(Ziegfeld 1926: 719)

The Penck/Fischer map proved to be very popular and slightly modified versions appeared in a variety of atlases[249] and other publications (see figure 7.3).[250] The map also initiated a new trend: its stress on the German cultural heritage in the East not only provided support for far-reaching

Das Deutschtum in Osteuropa.

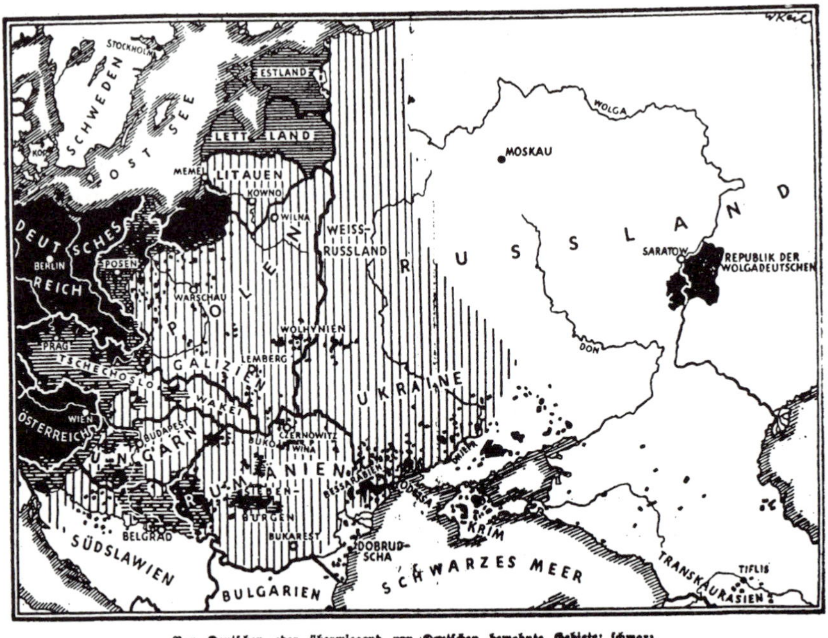

Von Deutschen oder überwiegend von Deutschen bewohnte Gebiete: schwarz, Gebiete deutscher Kultur: waagerecht gestrichelt. — Gebiete deutscher Verkehrssprache: senkrecht gestrichelt

Figure 7.3 Slightly modified reproduction of the Penck/Fischer map (Spohr ca. 1930: 32–3).
The different types of shading used in the map are described in the German caption as follows: Areas inhabited by Germans or inhabited mostly by Germans: black; areas of German culture: horizontal shading; areas of German language use in commerce: vertical shading.

territorial demands, but evoked a renewed sense of pride in German achievements. This sparked the production of maps on all aspects of German culture, as can be seen in an advertisement by the German atlas publisher Verlag Velhagen & Klasing:

> While regional maps are the necessary basis of every atlas, so to speak its roots, maps of German cultural heritage are its most noble adornment. . . . Despite the fierce competition in which our people is engaged for a place in the sun, such maps have only appeared recently and almost hesitatingly in atlases.
>
> (cited after Ziegfeld 1926: 715)

New maps stressing German culture, such as the distribution of cities with German law as far east as Kiev (Weizsäcker 1926: 553), became commonplace. The 72nd edition of the Diercke atlas in 1932 substituted the previously used map of the distribution of Europeans and Germans with a map which only showed the distribution of Germans and emphasized German settlements and the former German colonies. In the 50th edition of the Putzger atlas in 1931, the editor and publisher stated their desire that this new edition would "help form an historical awareness in our people and thus contribute to the renewed ascent of our fatherland."[251] To this end, the atlas now stressed German issues in an unprecedented manner: maps on the development of urban and rural settlements and postal services focused exclusively on German examples and two new maps on Germandom were added (Wolf 1978: 709). One of the maps, "Germandom in the world" (*Das Deutschtum in der Welt*), used areal coloring and depicted Germany and the former German colonies in blue, while the categories "German population numerous" and "German population sporadic" were shown in two shades of gray which covered most of Europe, stretching as far east as the Ural mountains, most of North America as well as large parts of South America and Australia. The other map was a slightly modified version of the Penck/Fischer map of the German *Volks- und Kulturboden*.

Organic unity and military threat

In the late 1920s, two other groups of maps emerged which reinforced the nationalist message of maps stressing German culture: geo-organic maps and depictions of the negative definition of national territory. This gave added support to revisions of the Treaty of Versailles: maps paying homage to the uniqueness of German culture justified expansion of German territory by virtue of its superior culture, geo-organic maps supplied economic arguments, and depictions of the negative definition offered strategic reasons. In all cases, the focus was toward the East.

Geo-organic unity maps, which were essentially based on economic

111

aspects, such as transportation networks or market areas, were first used in the early 1920s to argue for the indivisibility of Upper Silesia. They experienced a renaissance with the advent of the world economic crisis in 1929. Several new maps were published which stressed the disruption of the economic integrity of Germany's eastern areas by the new boundaries. They showed how the new boundaries dissected the existing transportation

Figure 7.4 Disruption of the geo-organic unity: dissected railroad lines (Ströhle 1931: 61).
The categories of the legend are (from top to bottom): LEFT: 1) German cities (1910) with more than 100,000 inhabitants; 2) German cities (1910) with 50–100,000 inhabitants. RIGHT: 1) German and Polish railroad lines, 1914; 2) Railroad lines built by Poland after the war. The same map, but without the lettering *"Polen"* appeared in: Der deutsche Osten, *Der Heimatdienst*, vol. 10, January 1930: title page.

systems, how economic market areas were closed off, and how railroad connections had been severed and reduced to a trickle (for an example, see figure 7.4). Some of these maps apparently were shown during an exhibition of the *Reichsverband der heimattreuen Ost- und Westpreußen* at the *Zentralinstitut für Erziehung und Unterricht*, Berlin, 8–14 January 1933, entitled *"Ostpreußen - was es leidet, was es leistet."*[252] Other maps contrasted the inorganic structure of the new economic system with the previously existing harmony. The economic structure in the eastern part of Germany before 1919 was portrayed as an organic unity of concentric circles with interconnecting arrows. After 1919, the economic condition of the same region was depicted as being completely disrupted by Polish competition.[253]

The primary intention behind the geo-organic unity maps was to rally support for the plight of Germany's eastern provinces, which were affected most by the general economic crisis. The majority of the maps were issued by institutions and authorities in the eastern provinces, such as *Landeshauptleute* (administrative heads of provinces) or university professors.[254]

Walter Geisler, professor of geography at the Technische Hochschule and the University in Breslau, Silesia, was particularly active. He published an atlas dedicated to the economic integrity of Silesia (Geisler 1932b) and founded a new monograph series in 1932. The goal of the series was "to address singular current problems of the German East in short studies, which are supported by maps whenever possible, and also to show the economic structure in its individual parts through the analysis of the spatial organism" (Geisler 1932a: preface). The first issue looked at Silesia as a spatial organism (Geisler 1932a) and the second issue addressed global economic aspects of the corridor and Upper Silesia (Werner 1932). Both used several maps to support their arguments. For example, one of the maps in the first issue showed how Germany's pre-war boundaries coincided with the division of two different cultural landscapes: settlement forms changed drastically along the old boundary. In the second issue, one map showed how the new political boundary dissected the industrial region of Upper Silesia, another depicted how this dissection affected the production and transport of Upper Silesian and Polish coal in 1913 and 1927.[255]

The geographer Erich Obst (1929: 758–9) addressed the detrimental effects of the border on the integrity of Upper Silesia in a similar way, as did publications by the *Reichszentrale für Heimatdienst* (Ulitzka 1932: 182). The common design of all these maps is the depiction of the industrial region as a dense cluster of factories, foundries, and mines which is arbitrarily divided by the new boundary.

Analogous to the paucity of publications on military geography (Kost 1988: 143–4), maps dealing with the strategic shortcomings of Germany's territory – that is, maps employing the negative definition of national

Die Bedrohung des deutschen Ostens und Südens

Figure 7.5 Threat to the German East and South (Boehm 1930: 309).

territory – received only little attention until the late 1920s and early 1930s. They were the exclusive domain of suggestive cartography and their focus was on the East, where Germany's boundaries were characterized by deep indentations and in the case of East Prussia even separation.

Common features were black semi-circles forming vices around exposed districts and arrows implying impending penetration. Even without indicating the actual strength of the neighboring countries' military, the graphic weight of the symbols made it clear that the situation for Germany was hopeless (see figure 7.5). Many of the maps highlighted the situation for individual regions, especially East Prussia and Upper Silesia. The most

Die Gefahrenlage Oberſchleſiens und Niederſchleſiens.

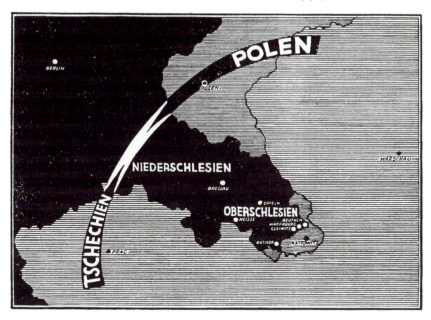

Figure 7.6 Silesia threatened by a pincer attack (*Reichszentrale für Heimatdienst*, Richtlinie Nr. 198/99, Grenzlandreihe 16, June 1930; BfPB).

widely disseminated one showed the threat to Silesia from a pincer attack: thick arrows originating in Poland and Czechoslovakia converged where Silesia connected with the rest of Germany and cut off the district. It appeared not only in the *Zeitschrift für Geopolitik* (Obst 1929), but also in a newspaper, the *General Anzeiger Bonn* (11 July 1930, p. 10), and in various publications by the *Reichszentrale* (see figure 7.6).

These maps were part of a general trend to challenge the armament restrictions of the Versailles Treaty in the late 1920s and early 1930s. The majority of publications used maps of heavily armed neighboring countries with their fortresses, armies, and military hardware amassed before the unprotected boundaries of Germany or maps showing the reach of foreign aircraft. Even the German government employed such maps to present its position more effectively at the 1931 disarmament conference in Paris (see figure 7.7).

CONFLICTING MESSAGES

Changes in the design of existing maps and the publication of new maps in the Weimar Republic resulted in a more favorable overall message for the

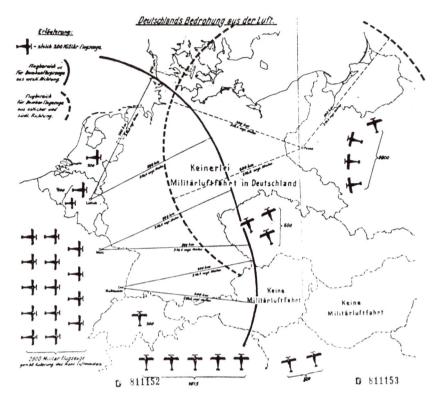

Figure 7.7 Germany's vulnerability to air attack (*Pressematerial des Reichswehrministeriums*; BA R431/518).

German cause during the later years of the Weimar Republic: the majority of maps supported further-reaching claims to German territory than before. Yet, by the mid-1920s there was still a considerable number of maps which supported the boundary delimitations of the Paris peace treaties.

Arnold Hillen Ziegfeld lamented in a 1926 survey on maps used in schools that some maps in standard textbooks in geography and history still had serious faults: one subsumed Masures and Kaschubes under the Poles, another showed a distribution of Germans in Upper Silesia which clearly had neglected the most recent research by Volz and his students (Ziegfeld 1926: 718).[256] Wilhelm Volz was even more disillusioned. He believed that there had been virtually no change by 1926:

It is common knowledge how much damage our "nationality maps" have done to the German cause, first in public opinion and then during the peace negotiations. Nonetheless, the lessons which arose out of this

116

have almost nowhere been applied in our atlases and, as a result, the detrimental effects continue.[257]

Volz's statement was certainly an exaggeration, given the alterations in designs of ethnographic maps in school atlases which could be observed in the mid-1920s. Nevertheless, even as late as the early 1930s, there was no uniform depiction of the extent of Germany's "true" national territory. In a report on the 24th Convention of German Geographers in Danzig (Gdansk) in 1931, the Westermann Verlag sales representative Herbert Heyde mentioned that he had heard repeated complaints about German publishers, who "even today, still supply the enemy unreflectively with material for his cause." He pointed out that maps from *Putzgers historischer Schulatlas* and from *Andrees Handatlas* were used to support Polish claims. They were cited in a brochure by Adam Thomas, entitled *Der polnische Korridor und der Friede*, which was published in Warsaw in 1930 with the assistance of the *Institut für Internationale Forschung*. Heyde had obtained a copy of the brochure from the *Polnische Aktion* in Danzig (Gdansk).[258] Similarly, a participant at a meeting of the VDA in June 1935 was able to illustrate the inconsistent depictions of German ethnographic and nationality maps with several examples.[259]

Conflicting messages existed even within the same atlas. In 1929, the 68th edition of the Diercke school atlas introduced a new ethnographic map of Poland on page 128a which showed a markedly different distribution than the ethnographic map of Germany on page 140. While the ethnographic map of Poland depicted the corridor and parts of Upper Silesia as mostly Polish (green), the ethnographic map of Germany chose cross-bars indicating a mixed Slavic/German (red/green) population. However, in the case of East Prussia the ethnographic map of Poland represented the entire province as German (red) whereas the ethnographic map of Germany used the German hue (red) only for the northern part and a mixed German/Slavic symbol for the south. The design of the ethnographic map of Poland remained the same until the 73rd edition of 1933, when the distribution for Upper Silesia was changed to resemble the ethnographic map of Germany. However, the depiction of the corridor and East Prussia continued to be different in the two maps.

These inconsistencies in design changes as well as the continued production of some new maps, which *völkisch* cartography experts, such as Ziegfeld (1925: 432–3), considered flawed and "unsatisfactory," reveal that the "cartographic Germanization" which took place during the Weimar Republic was a gradual, but by no means uniform, process. Government censorship or regulation, which would have ensured a unified message, was surprisingly absent. Instead, the government chose to give general support to private or semi-private institutions which were active in the mapping of German issues, such as the VDA and the *Stiftung*. At times, these organizations became so powerful that they even eclipsed the work of government bodies: the Weimar ministry of information, the *Reichszentrale für Heimatdienst*, wanted to publish a reference work on Germandom but gave up the plan when the *Stiftung* started a similar project.[260] In the case of the map *Deutsches Grenzland in Not*, the government did become active in regulation after repeated complaints from abroad, but it chose not to intervene very seriously. In 1929, the press division of the Foreign Office remonstrated with the *Deutsch-österreichischer Volksbund*, Berlin, which disseminated the map. When the organization continued to sell the map for a small fee, the Foreign Office simply contacted a member of the executive committee but apparently did not use any stronger measures.[261]

Maps became more Germanic in the mid-1920s because of the joint efforts of *völkisch* organizations and academic geographers who shared the same nationalist ideas. Many of the members of these two groups were influential and respectable figures with connections to the government and political parties, which enabled them to secure funding for their activities.[262] They founded institutions such as the *Stiftung*, developed new cartographic concepts, disseminated maps, and exerted pressure on map publishers through publicity campaigns. This stimulated nationalist pride and also instilled the fear of being chastised as "unpatriotic," or even of being accused of "treason." For example, Ziegfeld (1925: 445), in criticizing a map of the Czech language area included in a language guide by the Langenscheidt Verlag, declared that the publication of this map by a German publisher constituted "cartographic high treason" (*kartographischer Volksverrat*). To avoid such incidents, some publishers practiced self-censorship. Before publishing a map, they voluntarily submitted their works to the *Stiftung* or *völkisch* groups for examination.

The time period of the most marked change in German maps, the mid-1920s, and the focus of most maps on the East were influenced by general political developments. The economic prosperity of the "Golden Twenties" of the Weimar Republic restored German self-esteem and nourished feelings of nationalism, while the Treaty of Locarno of 1925 and the Russo-German Pact of Friendship of 1926 pointed revisionist efforts to Germany's eastern boundaries.[263] However, most of the map authors denied revisionist intentions.[264] They stressed that the changes reflected the fruits of objective research or that the new maps, such as *Deutsches Grenzland in Not*, simply showed the "cultural community of Germandom" (*Kulturgemeinschaft des Deutschtums*).[265]

8

CONCEPTS OF NATIONAL TERRITORY IN THE THIRD REICH

Like bombers and submarines, maps are indispensable instruments of war.

(Wright 1942: 527)

No troops have crossed the border without a map of the neighboring country; . . . If the distribution of Germans in these areas had not been recorded in maps, nobody would have had an idea of the German national territory (*Volksboden*).

(Welcoming speech, October 1938 meeting of the *Deutsche Kartographische Gesellschaft*)[266]

The replacement of Weimar's parliamentary democracy by the totalitarian system of National Socialism in 1933 fundamentally transformed Germany's political life. Yet, the change was not abrupt and was not just caused by coercion from above. Several elements of Hitler's Germany were already well established during the Weimar period, and voluntary submission to the Nazi ideology was widespread. This combination of continuity, coercion, and adaptation also applies to the mapping of German national territory.

MODIFICATION OF EXISTING CONCEPTS

In the Weimar Republic, maps of national territory initially comprised ethnographic maps, which employed language surveys or plebiscite results as criteria. When research revealed that not even the most accurate representations sufficed to claim what nationalists considered to be Germany's rightful territory, new maps appeared. In contrast to the social or person-based definition of ethnographic maps, these new cartographic depictions used spatial or geographic definitions: *Volks- und Kulturboden*, negative definition of territory, and geo-organic definition. These spatial concepts dominated the period from the mid-1920s to the mid-1930s, after which they became either obsolete or increasingly modified through the introduction of foe images and cultural or racial elements. Thus, with the

119

Nazi advent to power, German national territory again was outlined through person-based or social definitions. But in contrast to the original social definition using language, these later versions were explicitly German-oriented since they emphasized German racial and cultural superiority as well as threats to Germans from other peoples.

The negative definition of national territory

The spatial version of the negative concept stressed the vulnerability of Germany as a result of its new territorial shape as well as military restrictions outlined in the Treaty of Versailles. Maps employing this definition became prominent in the last years of the Weimar Republic. They continued to gain in importance in the early years of the Third Reich, when they served to justify the Nazi military build-up. They were not only widely disseminated in journals, pamphlets, and books, but also incorporated into the school curriculum (see figure 8.1).

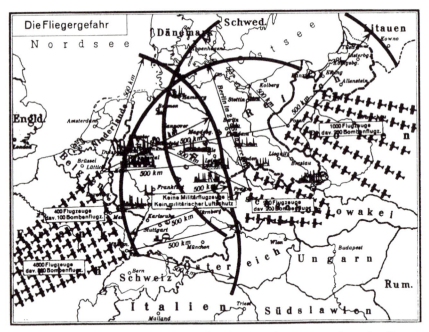

Figure 8.1 The negative definition of national territory: strategic and military vulnerability of Germany (Leers and Frenzel 1934: 40). [Note: black-and-white reproduction of a color original.]
The Leers and Frenzel atlas was used at a school of the *Landespolizei* and at the *3. Heeresfachschule* in Munich according to stamps in the copy at the University of Wisconsin-Madison library. A very similar map was included in *Putzgers Historischer Schulatlas*, 52nd edition, 1935.

However, while the depiction of Germany's strategic vulnerability was encouraged by the Nazis, the clear demarcation of the best strategic borders was not. The Nazis preferred to keep their options open and were careful not to appear overtly expansionist. As a result, they prohibited a book by the German geographer Ewald Banse (Banse 1932), which appeared during the last year of the Weimar Republic (Roessler 1990: 33). The book contained a map which outlined the "desirable" western military border of Germany as a line deep inside French territory: it stretched from Amiens to halfway between Reims and Verdun and continued west of Tull, Epinal, and Belfort (Banse 1932: 414).

After Germany had achieved full military sovereignty with official rearmament in 1935 and remilitarization of the Rhineland in 1936,[267] the spatial or strategic emphasis of the negative definition of territory lost in importance. Germany had ceased to be an easy target and the Nazi government had little interest in propagating such a view. On the contrary, the Nazis tried very hard to show that in only a few years they had made Germany powerful again and presented the restoration of German military sovereignty as one of their greatest achievements. Increasingly, the depictions of the threat facing Germany changed in emphasis: maps focusing on Germany's strategic vulnerability were replaced with representations which showed the hostile intentions of neighboring countries. Thus, by identifying a foe instead of a general danger, the threat became "personalized."

The transition from strategic vulnerability to threats from specific countries was gradual. As early as the last years of the Weimar Republic maps began to appear which shifted the emphasis of maps from strategic shortcomings to the source of the threat: a Slavic conspiracy (see figure 8.2).[268] The basis for these assertions of a pan-Slavic expansionism seems to have been a map which was published in the *Bohemian Review* in May 1918 (see figure 8.3). These more generalizing versions, which did not identify the exact source of the threat, but simply pointed to the Slavic East, soon became more specific: influenced by foreign-policy goals, maps began to target specific countries.

In preparation for the Saar plebiscite in 1935 and the invasion of France in 1940, maps addressed French aspirations to German territory or historic incursions. A map in a 1934 book on the Saar showed French plans from 1915 to reduce German territory to a small area which was to be subdivided into numerous principalities (*Das Saarbuch* 1934: 18).[269] The hostile intentions of the French were similarly illustrated in a map in a Nazi magazine from 1940 which depicted vast areas in Germany destroyed by the French in the seventeenth century (*Der Schulungsbrief*, 1940, 2. Folge: 2).

In the late 1930s, the emphasis was first on Czechoslovakia and then on Poland. The annexation of the Sudetenland and the establishment of the Protectorate of Bohemia-Moravia was prepared with maps which showed the territorial demands of Czech nationalists, such as Thomas Masaryk and

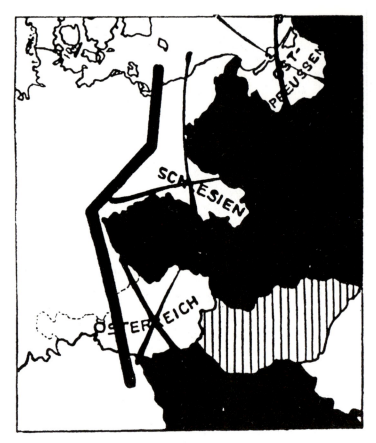

Figure 8.2 The Slavic conspiracy (Haushofer and Trampler 1931: 40).

Hanus Kuffner, and of the Czech delegation at the Paris peace conference in 1919.[270] Accompanying cartographic depictions offered "scientific" data to justify the "salvation" of the German population in Czechoslovakia: maps showed that unemployment rates, suicide rates, and the density of Czech nationalist organizations were highest in the Sudetenland.[271] Photographs of starving children (Heiss et al. 1938: Bildteil) and allegorical maps, which illustrated the Czech threat as a fist (see figure 9.4), added an emotional element.

After the annexation of the Sudetenland and the establishment of the protectorate, the focus shifted to Poland. This is most clearly revealed in two sets of maps; one contrasted Europe in 1914 with Europe in 1938, the other Europe in 1914 with Europe in 1939 (see figures 8.4 and 8.5). In figure 8.4, Czechoslovakia is the only country which is portrayed

Figure 8.3 The probable source for depictions of the Slavic conspiracy (*Bohemian Review,* vol. 11, May 1918: 71).

differently, and it appears as an intrusion into Europe, whereas in figure 8.5 Poland is shown as being disruptive to European unity.

In contrast to the anti-French and anti-Czech maps, which mostly used reproductions or redrawings of French and Czech maps from the time of

Die Mitte Europas im Jahre 1914

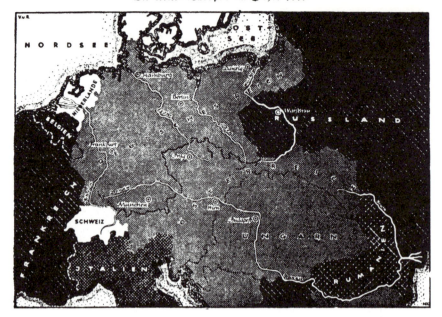

Die Mitte Europas im Jahre 1938

Figure 8.4 The center of Europe in 1914 and 1938 (Heiss et al. 1938: 296–7).

Die Mitte Europas im Jahre 1914

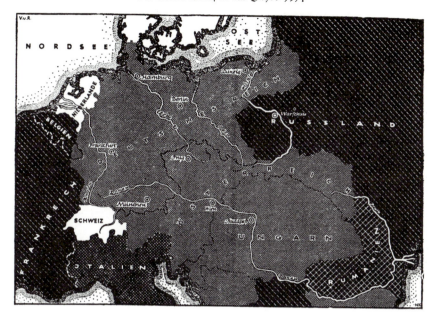

Die Mitte Europas im Jahre 1939

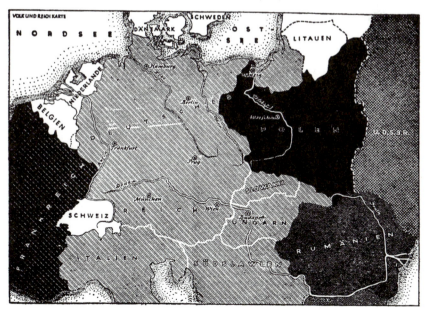

Figure 8.5 The center of Europe in 1914 and 1939 (Heiss et al. 1939: 296–7).

Polnifche Gebietsforderungen

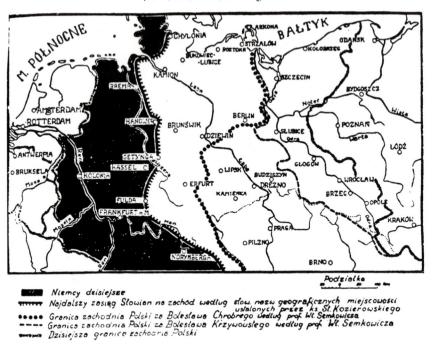

Figure 8.6 Polish aspirations to German territory I (Schadewaldt 1939b: 611).
A German caption below the map mentioned that the map had been published in
Dziennik Poznanski (no. 147) on 26 June 1939 and translated the Polish legend as
follows (from top to bottom): 1) present-day exclusively German settlement area;
2) Furthest west extension of Slavic territory after the geographical findings by St
Kozierowski; 3) Polish western border at the time of Boleslaw Chobres according
to professor Wl. Semkowicz; 4) present-day western border of Poland. [Note: the
German translation of the legend omitted one category from the original map.]
The map was also published in Heiss et al. (1939) and in *Facts in Review* vol. 2, no.
28, 8 July 1940: 294.

the First World War, the Polish threat to German territory was based on
maps which were presented as up-to-date Polish publications. For example,
in figure 8.6 the caption indicated that the original Polish map had been
published only a little while earlier, 26 June 1939, in the Polish newspaper
Dziennik Poznanski (no. 147). The postcard map in figure 8.7 was claimed to
have been distributed around the same time in Poland.

As in the case of Czechoslovakia, other maps gave "more scientific"
backing to the need to reclaim territories from Poland. Maps showed how
yields for agricultural products in the corridor area (e.g. wheat, sugar beets,
etc.) had suffered under the Polish economy (Schadewaldt 1939a). This was
"scientific proof" of the inability of Poles to run a country by themselves,

Figure 8.7 Polish aspirations to German territory II (Schadewaldt 1939b: 613). The postcard map was also reproduced and disseminated by the *Publikationsstelle-Berlin* in: Sommerliche polnische Wunschträume. *Polenberichte der Publikationsstelle*, no. 4, 12 June 1939 (AA Presseabt., Handakten Starke, Nr. 60).

an argument which had already been used in the cynical partitioning of Poland in the late 1700s and for which the Germans used the derogatory term *polnische Wirtschaft* (Burleigh 1988: 4).

Thus, the negative definition of territory – which entailed the general plea that more territory was needed to protect Germany adequately – was given added urgency by showing that the threat had a specific source: Germany was not just vulnerable, but an attack seemed certain given the aggressive tendencies of her neighbors. That this was precisely the message the Nazis wanted to give with such maps can be garnered from a plan of the 1943 exhibition *Kampf und Sieg des Nationalsozialismus*: "Maps, which document the annihilation plan of our enemies from time immemorial, show the inevitability of the war in which the German soldier has carried his banners deep into enemy territory" (BA NS 26/1395).

However, a clear government directive or unified stance pushing the dissemination and production of these personalized threat maps did not exist. As late as 1935, proposals for propaganda maps depicting Poland's imperialist tendencies were rejected by the Foreign Office as being in contradiction to the press agreement between Poland and Germany and therefore politically inopportune.[272]

The threat from neighboring countries was given additional support by a variety of related depictions. Maps showed that the birthrate of Germany had declined much more than that of her neighbors in the East between 1913 and 1933 (Korherr 1938: 31). This shifted the balance between German, Romance, and Slavic people in favor of the Slavs (Korherr 1938: 37), which made Germany vulnerable because the armies of her neighbors would grow at a faster rate than her own (Zu der Luth 1934: 7). Portrayals of earlier incursions from the East, such as invasions of Huns, Mongols, and Slavs, or Russian advances during the First World War, attempted to show that the present threat was not just an isolated incident, but part of a long tradition.[273] It was East against West, Slav against Teuton, Bolshevism against European culture.[274] Combined with illustrations of Germany's cultural legacy in the East (see figure 8.8), this elevated Germany's struggle to a struggle for the destiny of Europe (*Europas Schicksalskampf*).[275]

The new "personalized" emphasis provided a solid foundation for Nazi war plans. While it would have been very difficult to get the German people to go to war just to get less vulnerable boundaries, impending attack by Germany's enemies provided a more convincing justification. It made German expansion appear not just desirable, but a matter of life and death for the country, and for Europe. In addition, the Nazis could exploit the foreign threat for their population policies: a higher birthrate could be presented as a necessity for survival in view of the mounting population pressure from neighboring countries.

Once German military control over Europe was established after the victories against Poland and France, the negative definition of territory faded. The focus was now on depictions of military success, that is, on how the threat had been overcome, how the Nazis had "secured the Reich." For example, a pair of maps contrasted the secure location of Berlin in the center of the German empire in 1940 – it indicated that the distance to the

Figure 8.8 Eastern limits of German culture (Haushofer 1939: 110).

nearest borders was about 530 km – with the vulnerability of Berlin after Versailles when it was only 150 km from the Polish, Czechoslovak, and Baltic Sea boundaries (Lange 1940: 262–3).[276] The military even made movies with elaborate animated map sequences for this purpose (see Chapter 9). Shortly before the attack on the Soviet Union, some maps employing the negative definition reappeared, but they only achieved prominence again during the final stages of war, when maps of Allied plans to partition Germany into small regional fragments were used to appeal to the people to keep fighting until the very end.[277]

The geo-organic definition of national territory

During the Weimar Republic, the geo-organic concept, that is, the argument that the territories ceded by Germany in the Treaty of Versailles were an integral part of the German state organism, only had a notable impact in the beginning and the end. In the early 1920s, it served to argue for the indivisibility of Upper Silesia and as an inspiration for the development of the *Volks- und Kulturboden* concept. In the late 1920s and early 1930s it experienced a brief renaissance, because it provided a good argument for government support for Germany's border provinces, which had been badly affected by the worldwide economic recession (see Chapter 7, "The Emergence of New Maps").

After the Nazi advent to power, the geo-organic definition again lost in importance. Just as in the Weimar Republic, this was mainly the result of the restrictive character of the concept, which made it difficult to apply it convincingly to areas beyond pre-war Germany. Large ethnic German areas, such as the Sudetenland and Austria which had been part of a different economic unit (the Austro-Hungarian Empire), could not be reclaimed. This was as unacceptable to the Nazis as it had been to *völkisch* nationalists during the Weimar Republic. The Nazis had no interest in the restoration of the old boundaries; they wanted to create a new Greater Germany.[278]

In the first years of the Nazi reign, depictions of the economic *malaise* in the eastern provinces continued; they gave support to their argument that the Weimar Republic, which had accepted the Versailles Treaty, had led to the ruin of the country.[279] Afterwards, the concept was only applied on a regional level and very selectively. The most notable case was the Saar region plebiscite in 1935, which was to decide whether the region was to belong to Germany or to France. An atlas on the region, the *Saaratlas*, was published in 1934 which attempted to show that the region had always been "a part of the German state and economic body" (*Glied des deutschen Staats- und Wirtschaftskörper*) (Overbeck und Sante 1934: 13). Initially conceived in 1931 by a geographer and an archivist, the atlas, which appeared about half a year before the referendum, was used extensively in German propaganda.[280] It addressed a variety of economic, cultural, and historical issues to show how German influence dominated the region (Abel 1935: 539–40). For example, a series of maps depicted intensities in railroad commuter travel in 1929. They revealed how the commuter network extended beyond the Saar region into Germany but not into France (Overbeck and Sante 1934: 37).

Apart from the *Saaratlas*, the Nazis seem to have been ambivalent about regional applications of the geo-organic concept. In the spring of 1935, Walter Geisler conceived an atlas project about Upper Silesia which would show that the region was the product of German achievements and that the post-war boundaries had resulted in serious damage. He had difficulty in getting funding for the atlas and despite the glamorous portrayal in the preface, which listed noted scholars as collaborators, it appears to have been essentially a low-cost production based on the work of Geisler and some students.[281]

One possible reason for the Nazi ambivalence is that geo-organic maps stressed the decline in commerce in the affected areas. This was not very attractive to the Nazis, who tried to show that their economic policies and road-construction efforts (especially the *Autobahn*) had positive effects on even the most remote areas.[282] After Nazi Germany's conquest of Poland, which recaptured the eastern areas most frequently portrayed in geo-organic maps (Upper Silesia and the corridor), the geo-organic concept became obsolete.[283]

The *Volks- und Kulturboden* concept

This had been the most important definition of national territory during the Weimar Republic. It gained prominence because it was developed, applied, and propagated within a central institutional framework, the Leipzig *Stiftung für deutsche Volks- und Kulturbodenforschung*. When the *Stiftung* was dissolved in 1931, the concept was already very popular: the term was frequently used and maps of the *Volks- und Kulturboden* appeared in numerous publications. They were not only included in many atlases in the middle to late 1930s, but apparently even used on the back of a medical examination form for foreign students (Grimm 1939: 10–13). It seems that German geographers backed the dissemination of the maps with the weight of their professional reputation: a representative of the Westermann publishing firm who attended the meeting of the Association of German Geographers in 1934 (*25. Deutscher Geographentag*) claimed in a report that a resolution was accepted unanimously during the meeting which urged all publishers to include a map of the German *Volks- und Kulturboden* in all school atlases.[284] The popularity of *Volks- und Kulturboden* maps continued until shortly before the war, when Nazi institutions began restricting their publication.

However, the wide dissemination did not mean that there were no objections to the concept depicted on these maps. During the same meeting of geographers in 1934, Hermann Rüdiger cautioned that the term *Deutscher Kulturboden* had the potential for misinterpretation since it could give the impression that *Deutscher Kulturboden* was occupied by *Kulturdeutsche* (cultural Germans) (Rüdiger 1935: 94). The idea of "cultural Germans," that is, the possibility of other ethnic groups becoming German via cultural adaptation, was obviously incompatible with the racially defined notion of Germandom of National Socialism. In addition, the notion of *Deutscher Kulturboden* was too restrictive for the Nazis: it gave definite territorial boundaries to the area where German cultural influence was predominant. As can be seen from maps included in Nazi exhibitions and publications, the Nazis preferred more flexible cartographic depictions of German cultural influence, such as representations of the extent of German law in the East or the use of German language in commerce (see for example, figure 8.8).

In 1938, the tolerance of the Nazis for *Volks- und Kulturboden* maps wore

131

thin. A guideline of an NSDAP commission in charge of evaluating publications from November 1938 stated that while the achievements of Germans in the East should be emphasized, the term *Deutscher Kulturboden* had to be avoided.[285] Similarly, the Propaganda Ministry informed a publisher that the representation of the *Deutscher Kulturboden* in Europe was unacceptable.[286] However, the restrictions did not apply to "sections" of the *Deutscher Kulturboden*. Maps depicting *Deutscher Volks- und Kulturboden* in the East in combination with a category showing German cultural influence were used in the Diercke school atlas up to and including the 1940/1 edition.

THE EMERGENCE OF NEW CONCEPTS

Given the racial rhetoric of the National Socialists and the racial definition of who was German – which is exemplified by the 1935 Nuremberg laws on racial purity – one might expect a marked increase in maps of the distribution of races (*Rassenkarten*) during the Nazi regime.[287] However, only very few of these maps appeared (e.g. Folkers 1937). In fact, maps of races were strictly limited. In 1939, the *Parteiamtliche Prüfungskommission* prohibited the use of maps showing the distribution of races for primary school atlases.[288] And the Ministry of Education only allowed the map in H. K. F. Günther, *Rassenkunde des deutschen Volkes*,[289] for atlases for high schools (*höhere Lehranstalten*) and general geography textbooks (see figure 8.9). Maps of races for areas beyond Germany could only show the simple division into "white, yellow, and black."[290]

In 1941, the *Parteiamtliche Prüfungskommission* extended its censorship and maps of races were generally prohibited for school use (*grundsätzlich nicht zugelassen*).[291] The Nazis had good reasons for these restrictions: the distribution of Aryans did not extend beyond the eastern boundaries of Germany. Some maps by German racial scientists even showed racial boundaries which closely matched pre-war Polish claims (Burleigh 1988: 214).

In contrast to the limited use of maps of the contemporary distribution of races, depictions of the racial history, that is, of the pre-historic and Teutonic roots of the German *Volk*, experienced a surge in production. This increase lasted through the entire Nazi reign. But this did not reflect a general endorsement of historical maps of Germany's territory. In the National Socialist period, as in the Weimar Republic, representations of areas which were under German control during different historic periods, such as the Holy Roman Empire, were not very important. Historical maps were not useful in justifying German national self-determination, that is, a revision of the Treaty of Versailles. They could only be cited as supporting evidence to regain some regions previously occupied by Germans or to remind people of Germany's historical achievements.

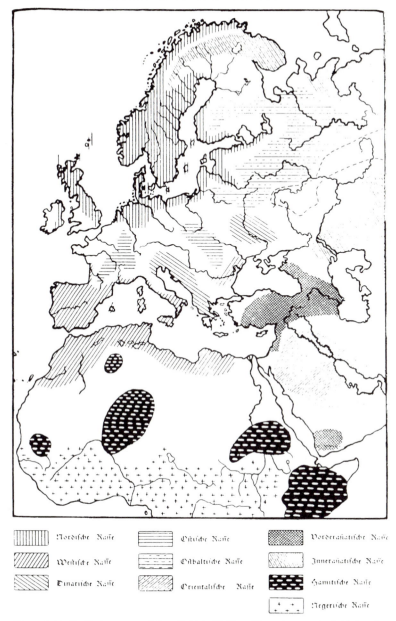

Die gestrichelte Linie umgrenzt das Gebiet stärksten Vorwiegens einer bisher nur ungenügend beschriebenen und meist „Pfafantypus" genannten Rasse; über diese vgl. „Rassenkunde Europas")

Karte XXI. Darstellung der Gebiete vermutlich stärksten Vorwiegens einzelner Rassen

Figure 8.9 The distribution of races in Europe and Northern Africa (Günther 1930: 307).
This is the only map in Günther's *Rassenkunde des deutschen Volkes* (1930), which gives an indication of the overall distribution of races in the German area.

Pre-historic and Teutonic maps served a different purpose for National Socialism. Nazi ideologues believed that history had a racial foundation, that race determined culture, and that the Nordic race, which was the origin of the Teutons, was the most superior. In this line of reasoning, the cradle of civilization was Central Europe, the home of the Nordic race and its Teutonic offspring (Rosenberg 1936: 205–7).

This Germanocentric view of history can also be seen in the inclusion of new maps in *Putzgers Historischer Schulatlas*. The 54th edition of 1937 introduced a map which presented Germany as "the heart of Europe" and as "the origin of the peasant peoples of the Nordic races." The map was contrasted with the "old" view of history which stressed the origins of urban culture in Mesopotamia and Egypt (Wolf 1978: 711).

The sketchiness and uncertainty of scientific evidence allowed for depictions which extended the Nordic and Teutonic influence all over Europe and toward the East. Maps were generally produced as series, which revealed the fluctuation of the extent of Nordic and Teutonic territory during different periods. They also employed arrows to symbolize far-reaching migrations. The fluctuations in combination with the arrows underlined the dynamic character of the ancient German territory.[292] This had the advantage of not limiting claims to a specific area, which left enough flexibility for Nazi conquests.

The shift in emphasis with regard to the concepts of national territory during the period of the Third Reich is revealed in the comparison of two editions of the *Volk und Reich* book *Deutschland und der Korridor*. The first edition (1933) was conceived in the last year of the Weimar Republic and authored by Friedrich Heiss and Arnold Hillen Ziegfeld, while the second edition (1939), which was completely revised, was prepared in close collaboration with the Foreign Office and authored by Friedrich Heiss, Günther Lohse, and Waldemar Wucher.[293] When compared to the 1933 edition, the 1939 edition had many more maps of German cultural heritage in the East, and more maps of the Teutons, but it also had fewer maps employing the geo-organic argument, fewer maps showing the extent of the German settlement areas (*Volksboden*), and no maps depicting military threats.

COORDINATION OF RESEARCH

With regard to research and development of cartographic depictions of German national territory, the end of the *Stiftung* in 1931 had meant the loss of a central coordinating forum. This left a void that could not be filled. Subsequent research and development activities dispersed into two main groups, the *Volkswissenschaftlicher Arbeitskreis* of the VDA, which was closely associated with the *Deutsches Ausland-Institut* (DAI) in Stuttgart, and the *Volksdeutsche Forschungsgemeinschaften*, which were made up of several regional subdivisions. Both groups held separate conferences and even though some researchers, such as Emil Meynen, participated in both (Roessler 1990: 124), the relationship between the two factions was strained (Ritter 1976: 130). After initial attempts at cooperation in the publication of a joint journal in 1935, disagreement increased and by 1937 two new journals were founded, the *Auslandsdeutsche Volksforschung*, which was edited by Hans Joachim Beyer of the DAI and the *Deutsches Archiv für Landes- und Volksforschung*, which was edited by members of the *Volksdeutsche Forschungsgemeinschaften* (Roessler 1990: 125).

The lack of integrating research through central conferences and publications had repercussions for actual mapping projects. At a meeting on 13 December 1935, which was convened to discuss the production of a map of the distribution of Germans for the *Atlas des deutschen Lebensraumes*, the editor of the atlas, Norbert Krebs, noted that only a few days previously, he had heard of a plan by the *Volksdeutsche Forschungsgemeinschaften* to make a similar map.[294] Yet, the plan by the *Forschungsgemeinschaften* had already been decided in March 1935, and the intention to include such a map in the atlas had been publicized in several reports since early 1934. During the meeting, which was a first attempt to coordinate efforts, it was decided that the distribution of Germans was to be based not just on official statistics but also on the results of local surveys. The map was to show the extent of German literary language use (*Schriftsprache*) and German national affiliation (*Volks- und Kulturbewußtsein*). As one of the participants of the meeting, Dr Papritz from the *Publikationsstelle-Berlin*, pointed out, a lot of preparatory research was already underway, but cooperation between the various institutions involved was still lacking.

A second meeting was held on 29 February 1936 in which it was suggested that Friedrich Metz of the *Zentralkommission für wissenschaftliche Landeskunde* – a scientific commission of the *Deutscher Geographentag*, which was to publish the map of the *Volksdeutsche Forschungsgemeinschaften* – be the central coordinator of the joint project.[295] It was mentioned at the meeting that it was necessary to collaborate with the VDA, but despite the ethnic German mapping activities of the VDA and DAI,[296] collaboration was not initiated. Contact was only established with an employee of the VDA, Franz Doubek, who already was associated with one of the *Volksdeutsche Forschungsgemeinschaften*, the *Nord- und Ostdeutsche Forschungsgemeinschaft* (NODFG).[297] Doubek had received a grant for RM 2,500 from the NODFG to prepare an atlas of German-language use in Poland.[298] He left the VDA in October 1936 and started to work full time at the *Publikationsstelle-Berlin*, which was associated with the NODFG.[299]

The fragmentation of research on German national territory went even beyond these two main groups because, increasingly, political groups became directly involved. The Nazi party's main office for political education (*Hauptschulungsamt*) intended to publish an edition of its series *Der*

Schulungsbrief in April 1938 which was to include maps and figures for the numbers of Germans in the world. Karl von Loesch prepared a detailed report which was discussed at two intensive, four-hour-long meetings held some months before publication. Among others, it included members of the *Volksdeutsche Mittelstelle* or *Vomi*,[300] the VDA (Lange and Fittbogen), SS (Lämmel), and *Bund Deutscher Osten* (Dorfmeister).[301] Yet, as Hermann Rüdiger pointed out in a letter to the *Amt für Schulungsbriefe*, neither the DAI, which possessed "very extensive material about Germandom abroad," nor himself, who was even listed in the meeting report as "most competent expert" (*fachkundigster Spezialist*), were directly consulted.[302]

Around the same time, early 1938, an attempt was finally made to achieve a complete integration of the mapping of German national territory: a series of meetings was held at the *Volksdeutsche Mittelstelle* (*Vomi*) which brought together the most important institutions and individuals previously involved in the production of such maps.[303] The first meeting appears to have been on 10 March 1938.[304]

The chief coordinator of the meetings, Horst Hoffmeyer of the *Vomi* and BDO, stressed that the goal of the sessions was to prepare a map of the distribution of Germans (*Deutschtum*) "on which all scientists agreed and . . . which was in our [the German] interest."[305] The map was to be used mainly by government institutions engaged in political education (*Schulung*) and was considered an urgent necessity.

The meetings were well structured to achieve the desired coordination of scientific and political issues: designated experts prepared working papers which formed the basis for discussion. Initially, the meetings were devoted to general and methodological questions; then the emphasis shifted to problems of individual boundary sections. The planned map was to depict the extent of the German *Volksboden*, but it was clear from the outset of the meetings that despite the popularity of the combination *Volks- und Kulturboden*, the word *Kulturboden* was not to be part of this concept. Even though they discussed what the term "German cultural landscape" entailed, the term *Kulturboden* was not considered as a category of the cartographic depiction. Instead, the map was to show the extent of three different categories of the German *Volksboden*. The National Socialist dislike for the term *deutscher Kulturboden* – a concept which stressed cultural rather than racial elements and which did not give much flexibility for territorial demands – had prevailed.

Not only was there a change in terminology, but the concept of *Volksboden* was modified to ensure that it could only be defined racially. The participants of the meetings eagerly incorporated anti-Semitic principles. There was general agreement that the German *Volksboden* had to exclude Jews because they were not rooted in the soil (*nicht bodenständig*), even if that meant that the extent of the German *Volksboden* was reduced in some areas. Only "true peoples" (*wirkliche Völker*), such as Germans, Poles,

Magyars, etc., were to be represented cartographically.[306] The concept had become explicitly anti-Semitic in character and Wilhelm Volz's more geographical version of *Volksboden*, which had already come under attack by *völkisch* geographers in the late 1920s (see Chapter 5, "The Demise of the *Stiftung*"), was now eliminated.

The shift to a *völkisch* or racial definition seems to have been initiated by Karl C. von Loesch, who had co-founded the Leipzig *Stiftung* with Volz, but who did not seem to have been actively involved in the earlier dispute between advocates of geographical and advocates of *völkisch* interpretations of *Volksboden*. Loesch vigorously pursued a racial definition of Germandom under the Nazis. He was one of the first experts on Germandom to have calculated figures for Germans "cleansed" (*bereinigt*) of German-speaking Jews. These figures had been used a few months earlier during preparatory discussions for the April 1938 edition of the *Schulungsbrief*.[307] He also made a study for the Academy for German Law (*Akademie für Deutsches Recht*) which dealt with terminology.[308] In it, Germans were defined as "all those Aryans, who belong to the German nation (*Volkstum*) and thus to the community of Germans (*deutsche Volksgemeinschaft*)" (cited after Roessler 1990: 66).

Loesch was not entirely satisfied with the outcome of the 22 April 1938 *Vomi* meeting and he subsequently prepared a paper dealing specifically with the cartographic depiction of the Jews. He argued that Jews who had assimilated in the various countries (*Assimilationsjuden*) should be treated "as if they did not exist," while the Jews living in ghettos in Eastern Europe had to be viewed as a people since their political importance was paramount and especially since their cartographic depiction ensured a reduction in the *Volksboden* of the Poles, Romanians, Ukrainians, and Lithuanians.[309]

The open support for a *völkisch* definition of *Volksboden*, that is, the strong emphasis on race, was part of the general ingratiation of scientists and others to the ideology and terminology of the new regime. This voluntary adjustment can also be seen in the emergence of maps after 1933 which used the term "blood" (*Blut*) to connote German ethnicity.[310] As late as 1925, Loesch still had rejected a racial foundation for the community of all Germans: "The Germans have anyway much more in common than a mere unity of language and 'culture' . . . a racial community in the sense of a genealogical similarity which can be scientifically proven . . . is however missing" (Loesch 1925: 16).

Yet, after anti-Semitic views and laws had become commonplace, Loesch not only adapted his views, but took part in their propagation. He was not a unique case. Many German geographers eagerly incorporated National Socialist ideology into their subject and put it at the service of the regime. Papers and remarks during their conferences, such as the 1934 *Deutscher Geographentag* in Bad Nauheim, stressed that geography meant "blood and soil" (*Blut und Boden*) and the relationship of race and space.[311]

The *Vomi* meetings in 1938 showed this voluntary submission of scientists to political goals even more clearly. When the participants recognized that the delimitation of the German core area with a 90 per cent majority of Germans would reduce claims in the East, they quickly agreed on a 80 per cent majority which was to be camouflaged with the description "80–100%" in the legend.[312] The scientists who participated in the meetings subordinated their expertise to political expediency: it was no longer accuracy or objectivity which determined the value of a map, but its usefulness to the German cause. Yet, the scientists still believed that they were objective and that they conducted scientific work. Two examples will illustrate this further.

Walter Geisler drew a map for one of the meetings and stressed in the accompanying paper that his intentions had been to make a cartographic representation which could be defended with scientific arguments, which was as close to reality as possible, and which was suggestive. At the same time, he admitted that his categories of the German *Volksboden* were specifically designed to allow claims beyond the existing state boundaries.[313]

In a working paper for the 30 May 1938 meeting, O. A. Isbert, the expert on the German–Magyar and German–Slovene boundaries in the southeast, presented a variety of German and foreign maps which could be used to determine ethnic boundaries.[314] These maps were not just large-scale representations in scientific publications,[315] but even included small-scale suggestive maps.[316] He discussed their benefits and shortcomings. Some depicted very clear ethnic boundaries, but limited German claims to a minimum; others expanded the German areas considerably, yet gave the impression that there was no unified German region because settlements were shown as dispersed enclaves. Isbert's paper made it very clear that a separate depiction of forest areas – one of the main demands of German scientists for ethnographic maps in the 1920s – was not in the German interest, because it fragmented the German–Magyar and German–Slovene border areas.

Precise definitions for what constituted German *Volksboden* were not given at the meetings, but the concept had clearly shifted away from cultural landscape features; it had become more based on *Volk* than on *Boden*, and *Volk* in this context meant race. For cartographic representation, three main categories were agreed upon during the *Vomi* meetings:[317]

Category I. *Geschlossener deutscher Volksboden* (German core area), which encompassed the contiguous German settlement area with an absolute majority of Germans based on language and other "ethnographic and cultural aspects" (*volkskundliche und kulturelle Momente*), excluding Jews. Initially, the German majority was defined with the 90 per cent figure, but for the Eastern border, the figure was reduced to 80 per cent. It is not clear whether the 80 per cent figure was also to be applied to the other borders.

Category II. *Übergangsgebiet* (mixed or transition area), which covered the zone between the German core area (category I) and the core areas of the neighboring peoples.

Category III. *Streudeutschtum* (scattered German area), which encompassed German enclaves inside the core area of other ethnic groups.

Despite the vague terminology, the participants were still able to delimit boundaries for the three categories. Guided by the premise of maximizing German claims, they set to work. They divided the German boundaries into regional sections and assigned experts to prepare papers and maps for discussion. After the group agreed on the best base map (which could be an existing map or one specifically prepared by the regional experts), the boundaries were outlined in detail. During this process it was determined whether more favorable recent data could be added, whether the data had to be restructured in different units, such as natural landscape divisions instead of administrative boundaries, or whether it had to be indicated in some cases that Germans had been forcibly expelled in order to make the German character of an area more convincing.[318]

It is not clear whether a final map based on this new definition of German *Volksboden* was ever drawn; I have found neither the sketch maps referred to during the meetings, nor a final version. However, a comparison of Penck's 1925 *Volks- und Kulturboden* map with the description of the boundaries in the incomplete meeting records offers the following glimpse of the new notion of German *Volksboden*:[319]

Northern, Western, and Southern Boundaries. There are no records of discussions involving the northern boundary with Denmark. The western boundaries of category I seem to have corresponded closely to the course of the German *Volksboden* outlined in Penck's 1925 map of the *Volks- und Kulturboden* (see figure 4.4). The course of the limit for category II was only described sketchily; it extended beyond category I in at least two areas: 1) from Sippenaken to Scherpensell, parallel to the political boundary with the Netherlands; and 2) in Alsace-Lorraine, where it reached about as far as the 1914 political boundary. Therefore, the *Vomi* meeting map made further-reaching claims than Penck's original version: in the West, Penck had not extended claims beyond the German *Volksboden*, because he believed that the German *Volksboden* and the German *Kulturboden* coincided there. The only exceptions were three small spots of *Kulturboden* in the area of Lorraine.

In the south, the boundaries of category I also corresponded to Penck's 1925 *Volksboden* limit, but along the German–Italian boundary, the representation of South Tyrol and Tarvis as German *Volksboden* was prohibited. The extent of category II in the south was not described in sufficient detail.

Eastern Boundaries. Category I of the Vomi meeting map extended far beyond Penck's *Volksboden*; in fact it covered much of Penck's *Kulturboden*. Category II was described very sketchily. It is not clear whether the areas of Penck's *Kulturboden* outside category I were to be claimed as category II.

East Prussia was to be represented in its entirety as category I; the Masurian minority in the south was to be omitted. The Memel territory also was to be category I, but the presence of Lithuanians was to be indicated with small crosses. In the corridor, the *Volksboden* was to be extended beyond the political boundaries as far as the line of demarcation (see figure 8.10),[320] and the decline of Germans in the region was to be indicated by symbols signifying the forcible expulsion of Germans (*Verdrängungskreuze*).[321] The limits of category II were not mentioned in the meeting notes.

In Upper Silesia, category I was to extend to the contemporary political boundary as well as the Hultschin district, and category II was to cover the entire part which had been ceded to Poland after the plebiscite. For Bohemia and Moravia, the notes are very sketchy. It was only mentioned that the delimitation of category II posed no problem, and that "the area between them is category II (transition), under no circumstance category III (scattered German settlements)."[322] This can only be interpreted to mean that category II was to encompass the entire area between Silesia and the northern parts of Austria, which Penck had designated as *Kulturboden* in his 1925 map, that is, the shaded area around the word "TSCHECHO-" in figure 4.4.

Based on the description, the *Vomi* meeting map would have offered few claims beyond Penck's version of the *Volks- und Kulturboden*. The main difference was that its argumentation was more in agreement with National Socialist ideology. It was based on an explicitly racial definition of Germandom and respected the German–Italian agreement on South Tyrol. Regardless of whether a final map was ever completed, its usefulness for public education (*Schulung*) or propaganda would have been limited: as early as 4 November 1938, the NSDAP commission in charge of evaluating publications, the *Parteiamtliche Prüfungskommission*, issued guidelines which prohibited overview depictions of the German *Volksboden*.[323]

The demise of the propagandistic value of the *Volksboden* concept did not extend to "applied" government use. For example, while work on the *Handwörterbuch des Grenz- und Auslandsdeutschtums* was officially ended on 1 April 1940 because of the changed political situation (Roessler 1990: 72), articles for restricted government use (*Vertraulicher Druck für Dienstgebrauch*) continued to be issued at least until 1944.[324] The reference work was

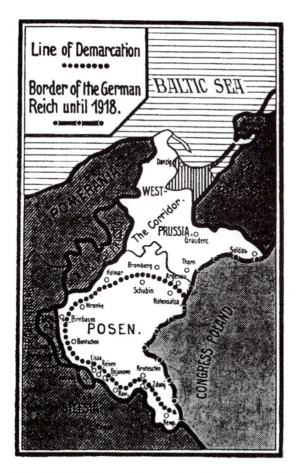

Figure 8.10 The line of demarcation (Werner 1932: 6).
The line of demarcation was the armistice boundary between Polish and
German troops which was decreed by the Allied and Associated Powers on
16 February 1919.

prepared in alphabetical order and by March 1944 the letters A through M
were completed. However, regions with high political priority, such as the
Netherlands, Slovakia, Hungary, and Turkey, had received preferential
earlier treatment.[325]

The case of the *Handwörterbuch* is indicative of the general shift in the
mapping of German national territory toward application for government
use which occurred in the late 1930s, when Nazi Germany started to
realize its expansionist goals. Up to that time, the propagation of the
newly developed concepts of national territory had served an important
function: it had given their far-reaching demands a cloak of scientific

respectability. Now, the Nazis were ready to carry out annexations and conquests and for these purposes they needed new maps. They no longer required maps which would maximize their claims regardless of the sketchiness of the evidence, but maps which were highly reliable and accurate. This meant ethnographic maps, since language surveys were still the most readily available and authoritative source of data.

A RENEWED INTEREST IN ACCURACY

As a result of the research on improved ethnographic maps during the early 1920s, German map-makers had acquired a keen understanding of the advantages and disadvantages of the different representation methods as well as a sound knowledge of the available data. Now, these lessons could be applied to cartographic production which emphasized accuracy. Ethnographic maps became actual tools for German conquest. They served several functions: the formulation and execution of foreign policy, the preparation of military operations, and the exercise of control in occupied territories.

Formulation and execution of foreign policy

While Austria (March 1938) and the Memel territory (March 1939) could simply be incorporated in the German Reich as existing territorial units, claims to German lands in Poland and Czechoslovakia had to be outlined with the help of ethnographic maps. Given the professed interest of the Nazi government in questions of German ethnicity and the unification of all Germans in one state, ethnographic maps provided a necessary tool for locating German ethnic settlements and for formulating policies toward that goal. In addition, reference to scientific ethnographic maps gave credibility to the official German position during diplomatic negotiations.

In the case of the German parts of Czechoslovakia, the so-called Sudetenland, the German government could draw on a map by Erwin Winkler, director of the Institute for the Statistics of Minorities (*Institut für Statistik der Minderheitenvölker*) at the University of Vienna.[326] Shortly before the negotiations between Italy, Germany, France, and Britain were to decide the fate of Czechoslovakia in Munich (29 September 1938), a slightly modified version of the map was published in the semi-official journal *Deutsches Archiv für Landes- und Volksforschung*.[327] The journal was edited by the directors of some of the regional *Volksdeutsche Forschungsgemeinschaften* (Albert Brackmann, Hugo Hassinger, Friedrich Metz), which had close connections to the Interior Ministry.[328]

Winkler's map served the propagation of German demands well because it had a very effective color scheme: all areas with 20–100 per cent German-language use were depicted by shades of orange or red, while the category 0–20 per cent was left white, which gave the impression that all areas with more than 20 per cent German-language use should belong to Germany. Through the publication of the map in a scientific journal the German demands were not only made known, but at the same time received scientific justification and blessing.

It is not clear whether Winkler's original 1936 map was made at the request of the German government, but in the 1920s, Winkler had collaborated with the *Stiftung für deutsche Volks- und Kulturbodenforschung* in Leipzig and received financial support from the Ministry of the Interior and the Foreign Office for some of his work.[329] He also made a map of the communal boundaries of the Sudetenland which was used by the German topographic survey, the *Reichsamt für Landesaufnahme*, as a base map for the boundary delimitations following the Munich Conference (29 September 1938).[330] The map of the communal boundaries had been commissioned by Professor E. Gierach, the founder of the *Anstalt für sudetendeutsche Heimatforschung* in Reichenberg, which was supported by the German Foreign Office through the Leipzig *Stiftung*.[331]

To complement Winkler's map, which depicted German settlements in the western parts of Czechoslovakia, ethnographic mapping was also extended to the eastern parts of the country. During a meeting of the *Volkswissenschaftlicher Arbeitskreis* of the VDA in January 1939 it was mentioned that surveys had been undertaken in Slovakia and Carpatho-Ukraine since 1936, because no reliable cartographic record existed for these areas. The materials were submitted at the request of the government in the fall of 1938.[332]

The distribution of Germans in Poland received special attention by government-affiliated research institutes. In early 1938, a map of the distribution of Germans in Central Poland appeared in the journal *Jomsburg* (Doubek and Horn 1938). The journal was subsidized by the Ministry of the Interior and the Foreign Office and published in close collaboration with institutions engaged in German national issues (Burleigh 1988: 140). The map, which used Penck's graduated version of a dot map (i.e., his absolute representation method described in Chapter 3), was prepared by Franz Doubek and Erna Horn of the *Publikationsstelle-Berlin*. It showed that Germans dominated even in areas beyond the pre-Versailles political boundaries of Germany, particularly around Lodz.[333] Thus, it gave "scientific" justification to the German occupation of Poland which was to follow.

In late 1938, a map series appeared in the journal *Jomsburg*. It depicted the national make-up of the districts of Freistadt and Teschen (Olsa-Schlesien), which Czechoslovakia had just ceded to Poland. The maps – which were again authored by Franz Doubek of the *Publikationsstelle-Berlin*, aimed "to show the weakness of Polish claims to Olsa-Schlesien" and to prove that "the only just allocation of this region would and should be to the German

Reich via the Protectorate."[334] The *Bund Deutscher Osten* criticized the maps for not being sufficiently pro-German, but the *Publikationsstelle-Berlin* stressed that it was necessary to represent the existing census figures accurately to avoid foreign criticism.[335] The map series was an outgrowth of work completed by Doubek for government use. At the order of the Foreign Office, Doubek drew ten large-scale maps (1:100,000) of the ethnographic situation in Olsa-Schlesien which provided the Foreign Office with detailed material for the formulation of policies for this Polish region.[336]

It is certain that even Adolf Hitler used an ethnographic map of Poland to "orient himself with regard to the Polish question."[337] The classified map,[338] which was "the first accurate and true cartographic record of the German *Volkstum* in the national structure of Poland,"[339] was based on a 37-sheet map series at a scale of 1:300,000. It depicted not only Germans, but also other nationalities, including Jews, which were prominently depicted in yellow – an indication of what was on the mind of the Nazis. Other users included the High Command of the Army, the General Staff of the Air Force, and the German Minister of Finance.[340]

At the end of June 1939, the NSDAP *Gauleitung Schlesien* requested "urgently needed" ethnographic material from the *Publikationsstelle-Berlin*.[341] They were interested in the quota of Germans in the Polish districts of Poznan and Silesia based on communal census data from 1910. There are no records of a response, but classified maps by the *Publikationsstelle-Berlin* have been preserved which depict the same type of information for a different Polish area.[342]

Preparation of military operations

The precise location of ethnic German groups (or other ethnic minorities which were pro-German) was indispensable for the military campaigns of Nazi Germany. The groups were potential collaborators and support bases and it was necessary to protect them from German fire, such as bombing raids or artillery attacks. The German military employed ethnographic maps not only in planning but also while carrying out operations.

In 1938, the military geographical service of the General Staff of the German Army issued a reduced version of Winkler's 1936 map, *Nationalitätenkarte der Sudetenländer*, which was restricted to "official use only."[343] This small-scale map was convenient for military planning and quick orientation. For the military occupation of the Sudetenland, the map was supplemented by a detailed 1:200,000 map of the Czech–German ethnic border prepared by the *Deutsches Ausland-Institut* in Stuttgart at the request of the *Wehrmacht*.[344]

For the attack on Poland, the German military used the same ethnographic map by the *Publikationsstelle-Berlin* which Hitler had consulted. In fact, the very first order for the map had been placed by the General Staff of the Air Force on 3 July 1939.[345] In August 1939, the High Command ordered 5,500 copies of the map, which also had an index on the back listing all settlements in Poland with at least 10 per cent Germans.[346] The map was successfully employed immediately before, during, and even after the actual military campaign.[347] Apparently, there was even an ethnographic map of Poland specifically designed for military warfare. It was entitled "*Wehrethnographische Karte von Polen*" and mentioned in a letter by the *Publikationsstelle-Berlin* to the German High Command.[348]

Judging from the lack of military orders for ethnographic maps on Western Europe, operations against France and the Benelux countries did not employ these aids. It appears that the more homogeneous national structure of Western Europe made ethnographic maps less useful in these areas. By contrast, the preparations for the attack on the Soviet Union and the campaigns in Southeastern Europe sparked the production of a number of ethnographic maps and maps of German settlements.[349] The different needs of the military operations in Eastern and Western Europe with regard to ethnographic maps can also be seen in military geographical atlases: Oskar Ritter von Niedermayer's *Wehrgeographischer Atlas der Union der Sozialistischen Sowjetrepubliken* (Niedermayer 1941) contained two ethnographic maps, while his *Wehrgeographischer Atlas von Frankreich* (Niedermayer 1939) had none.

Exercise of control

With the extension of German power over vast areas of non-German territory, scientists and politicians alike realized the importance of ethnographic maps for the "administration" of the occupied territories, that is, for their transformation into a Germanized territory. Himmler pointed out in a secret memorandum to Hitler (25 May 1940) that the knowledge of the distribution of ethnic groups in Eastern Europe enabled Germany to disrupt the national unity of Eastern European countries; they could be played off against each other (Burleigh 1988: 217).

During a meeting of the *Volksdeutsche Forschungsgemeinschaften* in Prague, the director of the central office (*Geschäftsstelle*), Emil Meynen, pointed out that the ethnographic maps prepared by the *Publikationsstellen* for Southeastern Europe were often "the only basis for the political structuring (*politische Gestaltung*) of this ethnically very intermixed (*verzahnt*) area by the armed forces and other official institutions." In fact, the work of the *Forschungsgemeinschaften* on ethnicity had even resulted in "a noteworthy increase in the direct involvement of foreign ethnic groups in the German service."[350]

An important task for accurate ethnographic mapping emerged with the attempt by the Nazi government to consolidate the ethnic German territory in Central Europe through resettlement schemes. The *Volksdeutsche Mittelstelle* received maps of the German settlements in the Soviet zone of interest in Galicia, Volhynia, and Northeast Poland from the *Publikationsstelle-Berlin*. At a scale of 1:100,000, they represented the percentages of Germans in the different settlements.[351] The actual resettlement of these ethnic Germans inside the German territory was also recorded cartographically. For example, the SS *Reichskommissariat für die Festigung des deutschen Volkstums* commissioned maps of the resettlement of Volhynian Germans in the Warthegau from the *Publikationsstelle-Berlin*.[352] According to Franz Doubek from the *Publikationsstelle-Berlin*, these population transfers to formerly Polish territory were aided by the index of German settlements on the back of the ethnographic map of Poland, which had been used during the campaign against Poland.[353]

The goal of the Nazis to expand and racially homogenize the German territory in Central Europe also directly involved non-German ethnic groups. Their fate ranged from expropriation and deportation to extermination. In January 1940, the Minister of the Interior distributed a map of the Slovak territory to the Foreign Office, the High Command of the Army, and the *Volksdeutsche Mittelstelle*.[354] It depicted the distribution of large estates by nationalities, which made it a convenient tool for organizing expropriation on the basis of the ethnic heritage of the landowners. The precise location of ethnic minorities inside Germany was also mapped to carry out Himmler's Germanization policies: for most of these groups, such as the Masures, Sorbs, and Kaschubes, the plan was to deport or exterminate the intelligentsia and to save the "racially valuable elements" (Burleigh 1988: 217). For example, for the Kaschubes, the SS *Reichskommissariat für die Festigung des deutschen Volkstums* ordered the production of a map which showed their precise distribution at the scale of 1:300,000; the map was completed in 1941.[355] The section for Racial and Ethnic Research at the *Institut für deutsche Ostarbeit* in Cracow – a Nazi research center founded by Hans Frank – prepared maps which showed the "relative ethnic-biological significance" of the ethnic groups in the *Generalgouvernement*. The purpose of these maps was to seek out the Germanic elements in the area (Burleigh 1988: 267).

The extermination of the Jews also seems to have been planned with the help of maps which applied the lessons on accuracy from the 1920s. Using the graduated version of the dot map developed by Penck, one map depicted the distribution of Jews in the East with graduated point symbols. They were grouped for the different localities to represent total figures accurately.[356] Another case has been reported which might have included cartographic depictions: in November 1939, the Gestapo received material from the *Publikationsstelle-Berlin*, which it had requested earlier. It showed the distribution of Jews in Polish cities with more than 10,000 inhabitants (Burleigh 1988: 165).

146

Data sources and designs of accurate maps

A variety of institutions were involved in the production of ethnographic maps to satisfy the "surprisingly increased demand" associated with wartime needs (Ritter 1976: 147). Most of the ethnographic mapping was carried out by institutes involved in research on ethnic German issues.[357] The most important institutions were: *Publikationsstelle-Berlin*, *Publikationsstelle-Wien*, *Publikationstelle-Ost* (*Sammlung Georg Leibbrandt*), *Deutsches Ausland-Institut*, and *Institut für Grenz- und Auslandsstudien*. They prepared maps not only for different government offices, but also for the military. Even though the German military had its own facility, the military geographical service, it lacked the experience and vast source materials of the specialized research institutes.

With accuracy as a prime concern of these applied maps, the institutes were eager to use the most up-to-date data. In addition to their own research and field work, they were kept abreast of foreign publications through the translation services of the *Publikationsstellen* in Berlin and Vienna, which were distributed regularly. Materials captured from libraries, university institutes, and archives in occupied territories often provided the only available data base for maps.[358] For example, the *Publikationsstelle-Wien* and the *Sammlung Georg Leibbrandt* made a classified map of the distribution of national groups in the Crimea which was based on a 1931 manuscript map by the Russian professor W. J. Filonenko which had been "secured" by a special SS commando in Simferopol.[359]

Direct contacts and exchange of materials with institutes in other countries which were involved in research on ethnicity (*Volkstumsfragen*) provided a further source of information. In March 1942, the director of the *Publikationsstelle-Wien*, SS *Hauptsturmführer* Krallert,[360] submitted a detailed report to the Foreign Office about the ethnographic maps produced at the *Staatswissenschaftliches Institut der Statistischen Gesellschaft* in Hungary. He mentioned that this facility was by far the most important he had visited in Hungary and that it had been founded by the former Hungarian Prime Minister Paul Teleki, who was also a geographer renowned for his innovative ethnographic maps. The institute was now under the directorship of Andreas Ronai, a student of Teleki. Krallert stressed that the existing close contacts to institutes in Romania – he did not mention their names – had to be extended to this important Hungarian facility.[361]

This emphasis on accurate and most up-to-date data and on cooperation did not extend to cartographic design. The various institutes did not address the issue of a unified design until 1944, when Krallert presented a plan for a uniform ethnographic map series to be prepared as a joint venture by the individual *Publikationsstellen*.[362] But the suggestion seems to have come too late; there are no records which would indicate that it was executed.

147

Most ethnographic maps used simple areal coloring, which had been severely criticized in the early 1920s for failing to show variations in population density. The only innovation from the 1920s which made a notable impact was Penck's modification of the dot map. Its greater accuracy was recognized and it was successfully employed in maps prepared by the *Publikationsstelle-Wien*.[363] The more complex innovative design by Volz, which had failed to attract much interest even when it was first presented in the Weimar Republic (see Chapter 3), apparently was only adopted abroad. Krallert saw a variation of the Volz method at the *Staatswissenschaftliches Institut* in Hungary and described its advantages in his report, but not a single German ethnographic map from the Nazi period with the Volz design is in the collections of the major archives.

A further example of design problems is a map made by the SS Security Service (SD) which constitutes a bastardization of Penck's dot-map method. The map depicted the ethnic composition of Latvia. For each community, the actual figures for the 1925 and 1935 census were inscribed for the categories "Germans," "Latvians," and "Others." German figures were written in red, Latvian in green, and Other in black. Cities were depicted slightly differently. But even though the total populations of the cities were crudely indicated with squares of varying sizes, no attempt was made to indicate graphically the percentages of the categories within the cities; only the absolute figures were written inside the squares.[364]

There are good reasons for this lack of attention to cartographic design. Many of the maps, particularly for military use, were rush jobs and the institutes involved in their preparation faced serious labor shortages. Sometimes the military waited until the last minute: at 10:30 p.m. on the evening before the attack on Poland, the military intelligence division (*Abwehrstelle*) of the German High Command called the *Publikationsstelle-Berlin* and requested information on the number of Germans in the Corridor area (*Pommerellen*).[365]

The demands on the institutes for the production of maps was extreme. For example, at the cartographic division of the *Publikationsstelle-Berlin*, which only numbered a few employees, an average of one new map was completed each work day in 1942.[366] Shortly before military campaigns the orders for ethnographic maps usually increased dramatically, but the work load could not be reduced by hiring new employees from other cartographic institutes given the general labor shortages during that time. In May 1939, the *Publikationsstelle-Berlin* requested three cartographers from the *Reichsamt für Landesaufnahme* to complete ethnographic maps but was informed that the *Reichsamt* itself was overburdened with urgent work for the German High Command.[367]

Given these time and labor constraints on production, areal coloring was the most efficient solution: most statistical data was listed by administrative districts and thus could easily be transferred to maps showing administrative boundaries.[368] Penck's graduated version of a dot map and Volz's method were considerably more time-consuming since they made it necessary to relate the statistical ethnographic data to settlement patterns and population densities. In addition, as mentioned in Chapter 3, Volz's method was not user-friendly: its uncommon and complex design was unintelligible without a descriptive legend. Its applicability for general use in the field or even for planning was therefore limited.

The complete absence of Volz's method in the Nazi period points to general differences in the concern for accurate ethnographic maps between the early 1920s and the late 1930s. In the early 1920s, their purpose was to serve as documents for a revision of the Paris peace treaties. The prime objective was scientific accuracy regardless of how easily it could be understood. In fact, the more complicated the method the better: this only underlined its scientific character. In the late 1930s, the purpose of ethnographic maps was to aid territorial conquest. The main goal was to produce maps which were not just as accurate as possible, but also as practical as possible. In the early 1920s, the concern for accuracy was academic; in the late 1930s it was applied.

THE DEMISE OF THE DEMAND FOR GERMAN SELF-DETERMINATION

The shift in the production of maps dealing with the issues of German nationality or ethnicity in the late 1930s toward applied government use also meant increasing restrictions on dissemination. Maps for the preparation of military operations and the exercise of control were secret (*nur für den Dienstgebrauch*), and maps used to justify foreign policy lost their importance with the extension of German control beyond the German *Volksboden*. For example, cartographic depictions of non-German ethnic groups, such as the Masures, Kaschubes, and Upper Silesians, on the fringes of the German *Volksboden*, became a classified matter.[369] Before the attack on Poland, the majority presence of these ethnic groups was used to argue that the cession of their territories to Poland in the Versailles Treaty violated the principle of national self-determination. With the establishment of German control over their territories this was no longer beneficial because it contradicted German claims.

Yet, the Nazis did not completely restrict maps of German national territory until 24 April 1942, when a decree was issued which prohibited maps dealing with German national or ethnic issues (*Volkstumskarten*) for the territory of the "Greater German Empire" (*Großdeutsches Reich*).[370] There was a secret press order on 29 July 1940, but it just stated that ethnographic maps of occupied territories "should be avoided" in order "not to anticipate possible future decisions of the *Führer*."[371]

It appears that the Nazi government waited until it believed victory was within its grasp before it decided to abandon the demand for German self-determination and to pursue alternative justifications for territorial claims. At the height of German military might, the Foreign Office initiated a massive mapping project, dubbed *Historisch-geographischer Atlas von Europa*, in July 1942.

The plan was to prepare an extensive three-volume collection of historical maps, contrasting the territorial development of Germany with that of other European countries. The atlas was intended to provide Hitler and the Foreign Minister with "valuable documents for the upcoming peace negotiations."[372] The dates of the maps were carefully chosen. For example, one map was to depict the political situation in the summer of 1918, when Germany briefly controlled vast areas in Eastern Europe at the end of the First World War. This would have shown that there was an historical precedent for the German area of occupation in 1942.[373]

The atlas was to be the best product German science could offer. Cooperation was sought from the most renowned geographers and historians and Emil Meynen, who had written the book *Deutschland und Deutsches Reich* (Meynen 1935), was designated as the editor in chief. Although numerous meetings were convened until the end of 1944, drafts for only twelve maps had been completed by that time.[374] The delay seems to have been caused by the changed military situation. A few months after the project had been started, Germany lost decisive battles, such as Stalingrad, and the Allied bombing raids on Berlin made it necessary to move the entire project from Berlin to Schloss Grabo in the fall of 1943; it was never completed.[375]

9

MAPS AND NAZI PROPAGANDA

The authoritarian style of government ensures that all publications of the book trade are geared to benefit the nation. . . . But there is no uniform line. The map publisher does not know which official institutions are interested in the works he is preparing, which demands they have, and how he can fulfill them.

(K. Frenzel 1938: 375)

It is generally believed and also propagated in the literature that Nazi propaganda was highly effective and well organized, not only during the actual Nazi reign, but also in the Weimar Republic. More recent studies, which place the examination of Nazi propaganda into a wider interpretative framework, show that the situation was more complex.[376] Although propaganda undoubtedly played an important role in bringing the Nazis to power, its success was not the result of superior organization or campaigns vis-à-vis the other parties, but of the "ability to combine the themes of traditional German nationalism with Nazi ideological motifs" (Welch 1993: 15–16).

For much of the 1920s, Nazi propaganda relied on emotive oratory by charismatic speakers and public spectacles in the form of torchlight processions, marches, or boycotts. The print media, film, and radio were dominated by *völkisch* nationalists, such as Hugenberg (Welch 1993: 12–13). This explains why Nazi propaganda based on graphic images only became unified and perfected after 1933, when the Nazis acquired control of the different forms of mass communication (Paul 1990).[377] However, this situation does not apply to cartographic images. National Socialist map propaganda did not achieve a uniform character even after the Nazi seizure of power and was plagued by irregularities and inconsistencies throughout its existence.

Differences between National Socialist propagandistic exploitation of images and maps can be traced to their historical origins. In the development of images for propaganda, the Nazis oriented themselves on their main adversary during the Weimar period, the German Left. The founding of the National Socialist satirical weekly, the *Brennessel*, in 1930, which was initiated by Hitler, was a direct response to the existing Left periodical *Simplicissimus* (Paul 1990: 145). Similarly, the Nazis modeled their illustrated periodical *Illustrierter Beobachter* and their general use of caricatures, photos, and images after communist examples (Paul 1990: 146–51). National Socialist propagandists, such as Erwin Schockel,[378] praised the effectiveness of communist posters, and Goebbels based his recommendations for National Socialist poster propaganda on the study of communist posters (Paul 1990: 150–1).[379]

By contrast, the propaganda use of cartographic images was pioneered by the German Right, in particular by people associated with the military or *völkisch* organizations (see Chapter 6). The German Left virtually ignored propaganda mapping. There seems to have been only one left-wing suggestive cartographer, Alex Rado, a native Hungarian, who was active in the communist party in Germany in the early 1920s. His *Atlas für Politik, Wirtschaft und Arbeiterbewegung* appeared in 1930,[380] when suggestive cartography had reached its peak and when numerous atlases were published by the Right. It seems likely that Rado's interest in suggestive maps was stimulated by his exposure to these suggestive maps in Germany and to the similar type of revisionist maps which were prevalent in Hungary. However, in the foreword to the atlas, Rado claims that his main inspiration came from a meeting with Lenin – obviously a more palatable explanation for a communist audience.

A closer examination of Rado's maps gives insights into remarkable similarities, but also significant differences, between propaganda cartography by the German Left and the German Right. Rado copied the style of suggestive cartography by using the same type of arrows and bold shading. Also, the distribution of Germans in his map "The Germans in Central Europe" shows substantial overlap with the minimum territorial demands of the German Right. Rado's map depicts less German territory in the corridor and Upper Silesia (see figure 9.1). This could be explained by the fact that sections of the German Left also rejected the Treaty of Versailles (Kershaw 1983a: 184). However, the majority of left-wing groups was opposed to nationalist aspirations; the international focus of communist ideology stressed allegiance along the lines of class, not nationality.

The captions accompanying this map and similar maps of the distribution of national groups in Rado's atlas resolve the contradiction between giving support to nationalist claims and the international ideology of communism (Rado 1980: 110, 122, 128). Rado included such maps of national territory not to support irredentist claims, but to expose the fact that the peace treaties at the end of the First World War used self-determination as a pretext to expand the power of the imperialist victors and to weaken the Soviet Union. Arrows identified as "zones of political

152

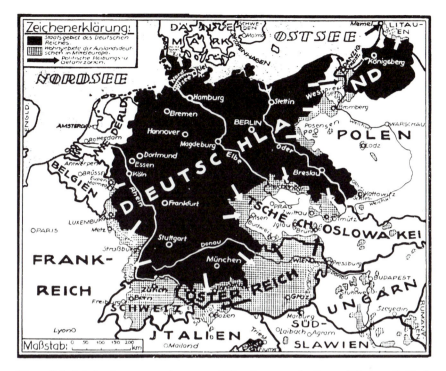

Figure 9.1 A left-wing suggestive map: "The Germans in Central Europe." (Rado 1930: 111).
The categories of the legend are (from top to bottom): 1) state territory of Germany; 2) settlement areas of the Germans outside Germany in Central Europe; 3) zones of political friction and danger.

friction and danger" showed that the new boundaries contained the seeds of future conflict (see figure 9.1). The solution was obvious: a Soviet-style communism which gave national groups autonomy.

The international focus of the German Left explains why their propaganda cartography was little developed. While German nationalists needed maps to define their goals for self-determination and to convince the public of the rightfulness of these demands, the Left did not have to outline specific areas: their claims were not limited to national territories or "regions of cultural influence," but extended to the "proletariat" of the entire globe.

Because suggestive mapping by the Left was insignificant and did not pose a threat, there was no need for the Nazis to become directly involved in cartographic propaganda; they could leave it in the hands of their political allies, who had mastered it. *Völkisch* nationalists, the military, and other right-wing groups shared many of the views of National Socialism and cooperated with the Nazis long before they publicly demonstrated their unity in a meeting at Bad Harzburg on 11 October 1931.[381]

153

Maps were nevertheless incorporated into National Socialist propaganda, but mainly those which left the options for territorial claims flexible and open. Depictions of threats to German territory and maps of the German heritage in the East were widely used as illustrations for publications. A special favorite were figurative maps designed to make their achievements look more impressive, such as the representation of the number of participants at the 1935 Nuremberg party convention as a line of people stretching from Berlin to Munich (*Illustrierter Beobachter* 1935: 1418).

Although the Nazis used maps, they thoroughly neglected their development: they did not pay attention to theory or terminology. This is doubly surprising since important strides were made with regard to theory and terminology in the early 1930s.

THEORETICAL ADVANCES

About a year after the Nazi advent to power, Rupert von Schumacher, an Austrian aristocrat, writer, and follower of *Geopolitik*,[382] offered the first theoretical treatment of design principles for suggestive maps (Schumacher 1934, 1935). In contrast to the other progenitors of suggestive mapping, who were trained in either geography or graphic design, Schumacher had a background in law. This explains why he never drew his own maps, but had two cartographers, Guido Gebhardt and G. Jedermann, execute his ideas.

In his 1934 article Schumacher offered a framework for the design of suggestive maps based on their audiences. He argued that scientifically trained readers were able to understand complex maps, which concentrated on the main issues without providing much background, but that the general public needed to be presented with highly simplified maps or map series (Schumacher 1934: 640–3). The effectiveness of this reductionist approach was illustrated with map comparisons.

Schumacher contrasted a scientific map of Germany's military situation with a popular depiction (see figure 9.2). The scientific map used four different types of line symbols, which included dotted, dashed, and solid lines. The design was elaborate enough to warrant a legend with a total of seven categories for military restrictions and target areas, such as demilitarized zones, areas where the construction of new fortifications was prohibited, industrial districts, and large cities. The popular version did not even have a legend and employed pictorial symbols to identify a smaller number of military restrictions and targets.

Another map comparison showed how a single issue – Germany's vulnerability to air attacks from neighboring countries – could be simplified to its most rudimentary elements to increase the propagandistic value of the map (Schumacher 1934: 643–5). Schumacher argued that the customary depiction of the threat from air strikes, which showed the range of foreign aircraft over German territory as overlapping semi-circles, was

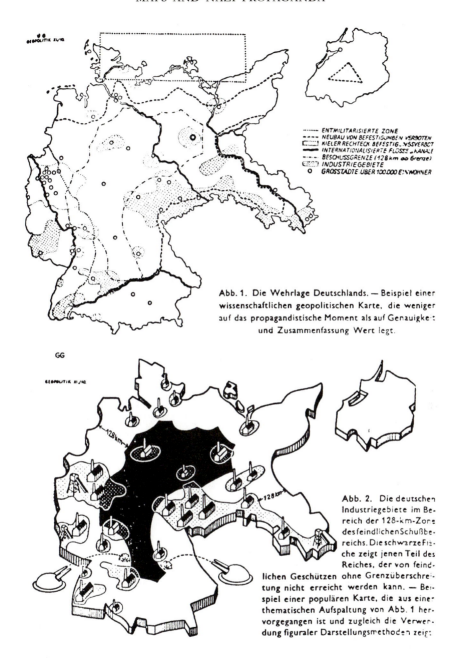

Abb. 1. Die Wehrlage Deutschlands. — Beispiel einer
wissenschaftlichen geopolitischen Karte, die weniger
auf das propagandistische Moment als auf Genauigke:t
und Zusammenfassung Wert legt.

Abb. 2. Die deutschen
Industriegebiete im Be-
reich der 128-km-Zone
des feindlichen Schußbe-
reichs. Die schwarze Flä-
che zeigt jenen Teil des
Reiches, der von feind-
lichen Geschützen ohne Grenzüberschre:-
tung nicht erreicht werden kann. — Beispiel einer populären Karte, die aus eine:
thematischen Aufspaltung von Abb. 1 her-
vorgegangen ist und zugleich die Verwen-
dung figuraler Darstellungsmethoden zeir:

Figure 9.2 Comparison between a scientific and a popular geopolitical map
(Schumacher 1934: 637).

155

not very convincing. A much more powerful design was to focus the threat from air strikes on a single country – Czechoslovakia – which gave the impression that even a minor state could destroy Germany (see figure 9.3).

However, Schumacher cautioned that the concentration on individual issues could easily lead to misinterpretation. He mentioned two possibilities: overly simplified maps and misleading comparisons. As an example of oversimplification, he presented a map of German territory with a circle drawn around it. In his view, the "encirclement" of Germany was not shown clearly in this manner and he suggested the use of thick arrows pointing at Germany from neighboring countries instead. This made the map "talk" by itself (Schumacher 1934: 644, 646).

With regard to misleading comparisons, Schumacher pointed out the detrimental effect of using two maps with differing designs. He showed representations of the continental position of Japan and Germany in two maps from different sources. The map of Japan used shading to designate the ocean, the map of Germany employed it to depict land, which was very confusing. Effective comparisons could only be achieved by absolutely identical designs. He supported his argument with a total of three maps:

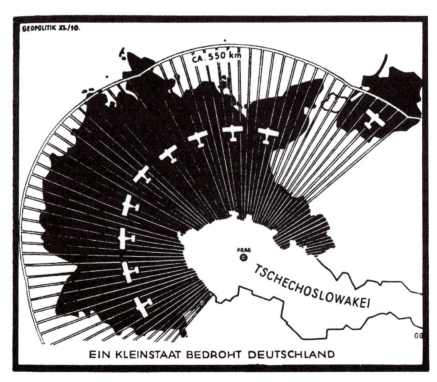

Figure 9.3 "A minor state threatens Germany" (Schumacher 1934: 645).

a map of Germany, a map of Japan with reversed shading, and a "corrected" map of Japan with shading that was identical to the German map (Schumacher 1934: 642).

Schumacher's reductionist approach, which gave suggestive maps an austere form (*strenge Form*) (1934: 649), included the use of lettering, legends, and color. The lettering had to be neat and clean, which meant that italics were not suitable. In addition, he asserted that the arrangement of the lettering should not make it necessary for the reader to turn the map in all directions to read it.[383] Legends should be avoided or at least integrated into the graphic depiction (Schumacher 1934: 650–1). As far as color was concerned, he rejected its unrestrained application and only advocated the sparing use of red or green (1934: 648–50).

Analogous to the geo-determinist emphasis of *Geopolitik*, which gave preeminence to the effects of natural landscape features on political situations, Schumacher recommended relief representations. His illustration was a map of Austria-Hungary in which mountain ranges were depicted with hachures. In his view this gave a much better understanding of the "causal connection" between the landscape and boundaries than the customary political maps devoid of such natural features. For popular maps the design could be improved even further by using the characteristic molehills of the early cartographers instead of more detailed shading or hachures (1934: 647–9).

In 1935, Schumacher extended his theoretical work to the symbology of suggestive maps and presented a catalog of 130 symbols classified in 11 subject headings which included topics such as attack, encirclement, and blockade.[384] He intended to provide guidelines for the use of suggestive symbols which, in contrast to regular cartographic symbols, were restricted to represent expressions of political power, such as the varying degree of power or the relationships between powers (1935: 251). Careful use of these symbols was necessary because the dynamic sign of the arrow in particular could always be interpreted by the enemy to express attack or imperialist tendencies (1935: 249–50). To prevent future misuse of symbols, he also included a catalog section of those he considered "nonsensical".

Schumacher's treatment of suggestive cartographic design theory was part of his larger interest in propaganda and political warfare. In 1937, he co-authored a book devoted to this subject, which also included examples of cartographic propaganda by other countries (Schumacher and Hummel 1937). Around the time of the Second World War, Schumacher commenced an ambitious research project on the legal and historical aspects of the Austrian military boundary. The ensuing voluminous book, *Des Reiches Hofzaun. Geschichte der deutschen Militärgrenze* (ca. 800 pages), found its way to the head of the SS, Heinrich Himmler. Despite a rather negative evaluation of the book by the *Reichssicherheitshauptamt*,[385] it impressed

Himmler, who ordered that Schumacher be employed by either the *Volks-deutsche Mittelstelle*, the *Stabshauptamt*, or the *Reichssicherheitshauptamt*.[386] Yet, it appears that no suitable position could be found for him.

Schumacher's theoretical work on suggestive map design was well received by other cartographers, who praised him for his efforts (Ziegfeld 1935: 243; Jantzen 1938: 320), but its general impact was limited. Even though Schumacher was a professed National Socialist who worked for some Nazi organizations,[387] his theoretical principles were not taken up by the government or the Nazi party.

There was only one incident in which the government acknowledged Schumacher's work. In 1938 and 1939, the Ministry of Education conducted a review of primary school atlases. Although the review mainly emphasized the representation of landscape, it also addressed general questions of design. One part of the review dealt with the misuse of arrows and the ministry recommended Schumacher's 1935 catalog of suggestive symbols.[388]

Apart from this brief reference, Schumacher's work was neglected, despite the fact that it could have served as a basis for standardizing suggestive mapping.[389] This had grave consequences. Walther Jantzen lamented in 1938 that most atlases and schoolbooks had serious design flaws because they did not heed Schumacher's principles (Jantzen 1938: 320–1). National Socialist publications suffered the most: except for maps prepared by the masters of suggestive cartography, Nazi maps had a less sophisticated design and violated many rules of suggestive cartographic theory (Herb 1989).

UNREFLECTIVE USE OF TERMINOLOGY

The Nazis not only failed to pay attention to theory; they also used terminology unreflectively. There was general agreement in the early 1930s among people involved in issues of German national territory that the German state should only be designated with the term *"Deutsches Reich"* and not with *"Deutschland."* The argument was that the term *"Deutschland"* should be reserved for the entirety of German national territory. This way, the goal to achieve political sovereignty over the entire German national territory could be given a terminological expression and summed up more succinctly and convincingly than with other words, such as *Volkssiedelgebiet*, *Sprachgebiet*, or *Volks- und Kulturboden*.

Völkisch nationalists and geographers considered this terminological distinction so important that they even sponsored the publication of a book solely devoted to the subject (Meynen 1935),[390] and the VDA vigorously attacked all maps which were labeled *"Deutschland"* but only showed the German state boundaries.[391] Yet, the Nazis continued to use the term *"Deutschland"* in this manner. For example, in 1936 a poster by the

Gaupropagandaamt Ost Hannover depicted a portrait of Hitler's head towering over the political boundaries of the German Reich together with the slogan *"Deutschland ist frei!"* (Germany is free!).[392]

The failure of the Nazis to recognize the value of this terminological distinction until the late 1930s is revealed in their ambivalent stance toward Meynen's detailed historical study of the concept *"Deutschland"* (Meynen 1935). On 19 February 1936, the *Reichsschrifttumskammer* prohibited the dissemination of Meynen's book, because it feared that it might give the impression abroad that Germany had expansionist tendencies. Barely five months had passed when the ban was lifted on 2 July 1936, because the Nazis realized that Meynen's work might be useful for them and that the restrictions were imposed too hastily. A committee had been formed under the leadership of the Ministry of Propaganda, which re-evaluated various book censorship initiatives.[393] The acceptance of the book was based on an expertise prepared by two German professors working at the *Reichserziehungskommission*. It argued:

> The . . . demand of the author not to limit the term *"Deutschland"* to the Reich territory, but to apply it to the contiguous German *Volksboden* . . . does not represent imperialism or annexationism . . . but only establishes the scientifically based fact of an old linguistic usage . . . How else could we sing in our German anthem *"von der Etsch bis an den Belt."*

> (BDC, personal file Emil Meynen)

However, the reversal of the ban only meant that the Nazis tolerated the book: the VDA was ordered not to draw foreign attention to the book through reviews in its journal. It took until the late 1930s before the Nazis fully comprehended the usefulness of the terminology and applied it for their purposes by purchasing a large number of copies of Meynen's book for political campaigning in Austria (Roessler 1990: 69).

ORGANIZATION OF PRODUCTION

Surprisingly, neither the government nor the Nazi party opted for the creation of a central cartographic propaganda institution. This institutional void was not the result of a lack of initiatives. One of the key developers of suggestive cartography, Arnold Ziegfeld, proposed the creation of a center for political cartography (*politische Kartenzentrale*) to the *Reichsministerium für Volksaufklärung und Propaganda* (Propaganda Ministry) on 3 April 1933, only three weeks after the ministry had been founded.[394] The center was to standardize regulation and to provide assistance for map production by various government authorities as well as to cooperate with private interest groups and the press.[395] Ziegfeld also attempted to get the support of the *Zentralinstitut für Erziehung und Unterricht*, where he gave a lecture on the

159

importance of suggestive maps for propaganda and education and sub-mitted a petition on 10 January 1934.[396]

Ziegfeld claimed later that a group of experts from all "applicable federal administrative offices" (*zuständigen Reichsbehörden*) fully accepted the proposed systematic development of suggestive cartography in 1933 (Ziegfeld 1935: 246). This could not be confirmed either in Ziegfeld's personal files at the Berlin Document Center or in the records of German government ministries. As it stands, some government officials might have agreed with Ziegfeld's proposal, but they certainly did not act on it: suggestive mapping continued to be a product of private enterprise.

In 1939, Ziegfeld made another attempt and voluntarily offered his services as a "specialist in modern political cartography" to the Ministry of Propaganda after the outbreak of the Second World War, in the fall of 1939.[397] Ziegfeld's initiative was fueled by his discontent with the position he was drafted into after mobilization: a clerk in an auto repair shop of the police (*Schutzpolizei*).

On 1 December 1939, he became a censor for maps in the German press in the Propaganda Ministry, but after one month he was transferred to the foreign propaganda office of the ministry, where he organized a map service for the foreign press and became involved in the production of the illustrated periodical *Signal*. During several months in 1940 and 1941, he went to the front as part of a propaganda unit (*Propaganda-Staffel*) and from October 1941 to July 1942 he was involved in directing the propaganda against England. On 31 July 1942, at the height of German military might, he submitted his resignation, arguing that he had to devote himself to his publishing firm, the Edwin Runge Verlag.

Ziegfeld's varied activities at the Propaganda Ministry reveal that the ministry did not take the opportunity to centralize cartographic propaganda after Ziegfeld had offered his considerable expertise. Not even the map service for the foreign press, which Ziegfeld had initiated, became incorporated into the Propaganda Ministry or other official Nazi institutions. The map service was only coordinated by the Propaganda Ministry: Ziegfeld commissioned the *Bibliographisches Institut* in Leipzig to prepare about four to six maps per week for use in the foreign press. The distribution was to occur via the Foreign Office, with which Ziegfeld collaborated.[398] The lax organization of the map service is indicative of the reluctance of the Nazis to become the main actor in cartographic propaganda. The Nazis preferred not to take an active part, but to cooperate with the established cartographers and institutions engaged in suggestive mapping.

COOPERATION WITH EXISTING INSTITUTIONS

Several key suggestive cartographers made maps for National Socialist party publications. In 1937, a map series by Karl Springenschmid, entitled

"Germany fights for Europe!" (*Deutschland kämpft für Europa!*), appeared in several issues of the Nazi party periodical *Der Schulungsbrief*. For the April 1938 issue of the same periodical, which was dedicated to the topic of Germandom, the party commissioned Arnold Ziegfeld to execute an oversized map supplement. The red and black map measured 21 × 14 in. (52 × 35 cm) and vividly portrayed the extension of German settlement (*Volksboden*) and culture far beyond the existing political boundaries. Friedrich Lange and Karl C. von Loesch had also collaborated on the map. Loesch had provided figures for the number of Germans in different regions which for the first time excluded German-speaking Jews.[399] However, apart from such individual contributions by suggestive cartographers, there are no records of Nazi institutions having established regular working relationships with these experts.

Another example of the tenuous links between Nazi party, government, and existing suggestive mapping institutions is the Marienwerder Office for Eastern Propaganda (*Amt für Ostwerbung*) in the *Bund Deutscher Osten*. The *Bund Deutscher Osten* (BDO) was an organization founded under the direction of the NSDAP, which united all *völkisch* groups involved in Eastern issues (*Ostverbände*) in order to streamline Eastern propaganda (Weissbecker 1983). The Marienwerder Office of the BDO was the successor to the *Arbeitsgemeinschaft für Grenzlandarbeit in Westpreussen*, which had disseminated corridor propaganda postcards and organized visits to the borderlands during the Weimar Republic (see Chapter 6). The office initiated the production of suggestive wall maps, which were to be used in its own propaganda efforts and those of the regional party leader school (*Kreisführerschule*), the *Hitler Jugend*, and the *Bund Deutscher Mädchen*.[400] Although the Propaganda Ministry supported the Marienwerder Office financially, it was not directly involved in the production of maps. Control apparently was only exerted through approval of items in the office's annual budget.[401]

There was only one institution which established a close and lasting cooperation with the Nazis: the *Stiftung Volk und Reich*. The foundation, which had become famous for its suggestive cartography already in the Weimar Republic, was able to offer its journal *Volk und Reich* and its monograph series as an inconspicuous medium for the dissemination of official propaganda. The impulse to establish close cooperation between the government and the *Stiftung Volk und Reich* came from foundation members.

Two members of the executive board of *Volk und Reich* pursued connections to the Foreign Office starting in March 1932. They were essentially motivated by difficulties in securing sufficient private funding because of the general economic malaise.[402] In July 1933, Friedrich Heiss, the director, made a further attempt and suggested to make the journal *Volk und Reich* into a tool of German foreign propaganda.[403] Heiss was successful: the publishing house Volk und Reich not only received support from the Foreign

Office, but ended up preparing basically the entire propagandistic literature for the attack on Poland in cooperation with the press office of the German Foreign Office.[404] For example, the August 1939 issue of *Volk und Reich*, entitled "War because of Poland?" (*Krieg wegen Polen?*), was published in English, French, Polish, and German. It contained several black and white maps and one colored fold-out suggestive map, and was apparently sent under the guise of being a special introductory issue to prospective subscribers to escape foreign censors.[405]

Volk und Reich also prepared maps for propaganda materials disseminated by the German Library of Information in the United States. The cover map for the 8 December 1939 edition of *Facts in Review*, which had been commissioned by the information office (*Informationsstelle*) of the Foreign Office, contrasted the successful encirclement of Germany in 1914 with the unsuccessful encirclement in 1939.[406] The same map was also used in German propaganda in Sweden: a total of 200 copies were sent to influential persons there on 16 November 1939.[407] In February 1940, *Volk und Reich* started to provide the Foreign Office with six to eight maps per month with English lettering for propaganda in the United States.[408] However, it is not clear if *Volk und Reich* was also involved in the production of the atlas *The War in Maps* (Wirsing 1941), which was disseminated through the German Library of Information, to convince the United States to remain neutral.[409]

It appears that after Heiss had succeeded in establishing a working relationship with the Foreign Office, he made sure that he would exercise complete control over the activities of the foundation *Volk und Reich*. On 12 August 1933, without notice, he fired Ziegfeld, who not only had established the suggestive cartographic office, but had also expanded the publishing activities of *Volk und Reich*. Ziegfeld had considerable expertise in publishing – he successfully ran his own publishing house on the side – as well as in issues relating to German territories in Eastern Europe. The termination of employment came as a complete surprise to Ziegfeld, who took *Volk und Reich* to court and requested an inquiry by the Foreign Office. In his letter to the Foreign Office Ziegfeld complained that Heiss had tried from the beginning to make Ziegfeld's contributions to *Volk und Reich* appear as insignificant as possible. The dispute was resolved with a financial settlement out of court, which also included the provision that Ziegfeld's name be mentioned if his maps or other work were used by *Volk und Reich* in the future.[410]

While Heiss's true reasons for dismissing Ziegfeld cannot be established, it is clear that the timing suited Heiss very well. The suggestive cartographic office was firmly established and its fame was associated with the name *Volk und Reich* and not with Ziegfeld. Most of the maps in *Volk und Reich* works only contained Ziegfeld's symbol, not his full name, and Ziegfeld was never given credit publicly for being the originator of the suggestive cartography of *Volk und Reich*. In addition, the contacts to the press office of the Foreign Office made Ziegfeld's publishing expertise less crucial. His scientific skills could also be replaced easily: in October 1933, Heiss contracted Karl C. von Loesch to assist in *Volk und Reich*.[411] At the time, Loesch worked at the *Institut für Grenz- und Auslandsstudien* and directed the *Seminar für Volkstumsfragen* at the *Hochschule für Politik* in Berlin.[412]

The *Volk und Reich* foundation had close connections not only to the Foreign Office, but also to *völkisch* organizations and Nazi party institutions. For example, *Volk und Reich* published a bibliography of works on German ethnicity and heritage, which had been prepared by the *Deutsches Ausland-Institut* in Stuttgart.[413] Walter Funk, a member of the board of trustees of the *Stiftung Volk und Reich*, who became press director and under-secretary of the Propaganda Ministry in 1933, was instrumental in securing substantial funds – more than RM 200,000 in 1934 – from the Propaganda Ministry.[414] *Volk und Reich* also made suggestive maps for the Hitler Youth and for several exhibitions.[415]

At the request of the *Amt für Weltanschauliche Schulung der NSDAP*, *Volk und Reich* produced forty enormous maps (3 m × 4 m) for the exhibit "*Kampf um Deutschlands Grösse*" in 1940. It also made seven wall maps for the book exhibit in Prag "*Auf dem Weg ins Reich*" (November 1941), as well as maps for the exhibit of the BDO, "*Der Osten, das deutsche Schicksalsland*" (4 December 1933–4 February 1934), and the 1938 party convention exhibit, "*Europas Schicksal im Osten*."[416] Other *Volk und Reich* publications were prepared for the *Reichsführer-SS*,[417] *Reichsminister für Bewaffnung und Munition, Generalinspektor für das deutsche Strassenwesen*, and *Generalbauinspektor für die Reichshauptstadt*.[418] Heiss constantly sought to expand the activities of *Volk und Reich,* and in August 1939 he suggested to the Foreign Office producing a propaganda film on the corridor which was to contain trick sequences with suggestive *Volk und Reich* maps.[419] However, there is no indication that his proposal was taken up.

In addition to cooperation between the Nazis and well-established suggestive cartographers and mapping institutions, there were also two new institutes which became involved in map propaganda: the *Publikationsstelle-Berlin (Puste)*[420] and the *Institut für Allgemeine Wehrlehre* at the University of Berlin. The *Puste*, following in the footsteps of the consulting work of the *Stiftung für deutsche Volks- und Kulturbodenforschung*, prepared reports and map drafts which ensured that suggestive maps still had a sufficient scientific basis not to be immediately exposed as false. The role of the *Institut für Allgemeine Wehrlehre* is somewhat obscure, it only seems to have been involved in one massive project: the map exhibit of the 1938 party convention.

The *Puste* was involved in the production of suggestive maps not on account of designs skills, like *Volk und Reich*, but because of its scientific expertise. Founded in 1932 on the initiative of Albert Brackmann, the director of the Prussian State Archives, the *Puste* was a central information office dedicated to coordinating and controling German research on the East as well as to responding to Polish and Czech scientific propaganda (Burleigh 1988: 51–9). The *Puste* and the affiliated *Nord- und Ostdeutsche Forschungsgemeinschaft* were closely associated with the Ministry of the Interior and had the same goal as the *Stiftung für deutsche Volks- und Kulturbodenforschung*: to supply scientific material for political purposes (Burleigh 1988: 53).

The *Puste* collaborated with several organizations, such as the BDO, the Nazi party, and the propaganda division of the German High Command.[421] For the BDO exhibit, "*Der Osten, das deutsche Schicksalsland*," the *Puste*'s expertise was used in the preparation of the historical section. The *Puste* was involved in the production of a map showing that the Polish ethnic boundary did not correspond with the present political boundaries but rather with the Polish territory established by Napoleon. It also worked on maps depicting Polish attacks on East Prussia in the fifteenth and seventeenth centuries.[422]

When the BDO wanted to publish a suggestive map of the Germans in the Sudeten area and a map of the German East, it requested assistance from members of the *Puste* and the *Nord- und Ostdeutsche Forschungsgemeinschaft*. A draft of the map on the Sudeten area, which was to be a wall map, was examined by Hermann Aubin, Walter Kuhn, and Franz Doubek.[423] Doubek, a specialist on German dialects in Poland, also submitted a sketch of the islands of German settlements located in the Polish part of the Sudetenland map. However, the experts requested that their names not be mentioned on the map in order not to compromise their future research activities. Doubek also sent a sketch map to aid the production of the BDO map of the German East and urged the BDO not to use a solid line to demarcate the limits of dispersed German areas in Poland.[424]

In preparation for the 1938 Nazi party convention, the *Puste* made drafts for several historical maps at the request of the *Amt für Schrifttumspflege der NSDAP (Hauptstelle II, Ausstellungswesen)*. They included a map of German migration to the East in the Middle Ages and in modern times, a map of the eastern boundaries of the Gothic building style, and a map of German commerce in Eastern Europe.[425] Most of the design work for the convention exhibit was carried out by *Volk und Reich* in collaboration with the *Puste*, but the *Institut für Allgemeine Wehrlehre* was placed in charge of coordinating the map production and executing the final wall maps in a uniform manner.[426]

The *Institut für Allgemeine Wehrlehre* was under the directorship of Oskar Ritter von Niedermayer, a German geographer and former employee of the *Reichswehrministerium* (Kost 1988: 175). In addition to organizing production and finalizing the maps, the institute was also entrusted with the production of a book on the exhibit. In response to numerous requests voiced during the exhibition, the central office of the Nazi party (*Reichsleitung*) decided to reproduce some of the maps and other materials in a publication which was to be used for political education (*Schulung*).[427]

Since a published work was more exposed to foreign criticism than the exhibit, Niedermayer asked all scientific collaborators, including the *Puste*, to re-examine the maps to make sure that they only contained material which was "undeniably scientific" (*nur wissenschaftlich einwandfrei Vertretbares*).[428] In response, the *Puste* withdrew one map on German commerce in Eastern Europe in the late Middle Ages because "the map was only intended as a one-time exhibition map" and could not be brought up to more rigorous standards without considerable work.[429]

Niedermayer's concern for the scientific basis of suggestive maps had a good reason. A Latvian newspaper had published a report which pointed out that one map in the exhibit with the title "German soldiers save the Baltic from Bolshevism" depicted a "Battle at Wenden" with the date 21 July 1919, that is, several months after the armistice. The newspaper had also reported that, according to the director of the exhibition, this map was a work by Niedermayer.[430]

REGULATION AND CONTROL

Since the Nazis did not create a centralized production network, one would expect that they tried to unify the political message of maps through *Gleichschaltung*, that is, the alignment of all institutions behind Nazi ideology, or through central censorship. Yet, even cartographic regulation was characterized by diversity and conflict.[431]

Up to the mid-1930s, official enforcement of cartographic messages was surprisingly light. Virtually no edicts or laws specifically addressed maps. This period was dominated by voluntary adaptation to the new ideology (*Selbstanpassung*) and by the control efforts of the alliance of geographers and *völkisch* nationalists which had developed in the Weimar Republic.[432]

The eagerness to adapt to the new ideology was a general phenomenon in geography and is revealed in statements voiced during the Conventions of German Geographers in 1934 and 1936 which stressed the importance of race and blood in geographical inquiry (see Chapter 8).[433] It also found expression in cartographic works. For example, Paul Diercke, who was working on a new edition of one of his school atlases, wrote to his publisher in 1933 that the treatment of races and German ethnicity and heritage had to be increased considerably in the atlas to pay tribute to the new political emphasis. This was not simply his personal view, but he could refer to studies which four authorities in the field – among them a professor of geography – had prepared at his request.[434]

Geographers and *völkisch* groups continued their public chastising campaigns from the Weimar Republic (see Chapter 7). The VDA stressed

165

in its publication, *Rolandblätter*, that the current German political territory should not be labeled "*Deutschland*," since Germany encompassed much more. It even reprinted advertisement maps which violated this principle.[435] Foreign ethnographic maps which were not deemed to be in the German national interest were also publicly scrutinized; for example, if they failed to distinguish Masures, Kaschubes, and Upper Silesians from the Poles (Geisler 1934; Eine unmögliche . . . 1938: 88).[436]

The geographer Friedrich Metz wrote a scathing review in the VDA publication *Deutsche Arbeit* (Metz 1934) about a German map which not only had identified the German political territory with the term "*Deutschland*," but also had used foreign place names for former German settlements.[437] The article reminded the reader of the Ullstein atlas incident in the mid-1920s which had led to the 1925 resolution by the German *Geographentag* to give preference to German place names (see Chapter 7). Metz argued that this resolution and the publication of German place-name dictionaries by the *Zentralkommission für wissenschaftliche Landeskunde* (Gradmann 1929) and by the *Deutsche Akademie* (Kredel and Thierfelder 1931) had given clear guidelines and that violations could no longer be excused. Therefore, he considered the publication of this map, especially after the "national uprising" (*nationale Erhebung*), to be an expression of "a lack of national dignity" (*nationale[r] Würdelosigkeit*). During the *Geographentag* in 1934 in Bad Nauheim, Metz, who was the director of the *Zentralkommission*, brought the Gea map to the attention of his fellow geographers (*Verhandlungen . . . 1935: 13*).

By the mid-1930s, these control efforts of the alliance between geographers and *völkisch* nationalists were increasingly supplemented by official activities. In October 1936 and April 1937, the Ministry of the Interior publicly embraced the 1925 place-name resolution of the *Geographentag* when it issued decrees which demanded that preference be given to German place names.[438] The *Publikationsstellen*, in particular the one in Berlin (*Puste*), acquired the role of advisory committees (*Gutachterstellen*) and prepared map evaluations and recommendations which were passed on to the Ministry of the Interior and the Propaganda Ministry for further action. However, cartographic regulation was not one of the original intentions of the *Publikationsstellen*; they became involved in this process rather by accident and they continued to carry it out in a haphazard manner.[439]

The first record of a map critique by the *Puste* is from October 1936, when the *Puste* happened to notice flaws in a BDO periodical which contained maps on Poland.[440] The *Puste* thought that the maps supported Polish claims and recommended their immediate withdrawal.[441] In particular, the *Puste* rejected the map of a unified Slavic territory in the ninth century, which reached as far west as the Elbe river, by arguing that there were great differences between the various Slavic groups. In another map it opposed the inclusion of Poland's 1772 border because it showed that up to that time Poland included the corridor, parts of East Prussia, Silesia, Lithuania, Latvia, and the Soviet Union. It pointed out that this historic boundary of Poland was frequently used by Polish nationalists in support of an expansion of Polish territory. As a general criticism, the *Puste* insisted that an atlas which purported to represent German interests had to show the Germanic pre-history of the region and devote at least four to five maps to this topic to give a visual impression of the long history of German settlement.[442]

The haphazard nature of the *Puste*'s control efforts become fully apparent in the following incident. On 30 July 1937, the *Puste* contacted the editorial board of the *Zeitschrift für Geopolitik*, to complain about a map in the June issue which showed the contemporary Polish state as a natural region.[443] The complaint by the *Puste* came several weeks after the map was published because the *Puste* only had become aware of the map through a positive review in the Polish government publication *Gazeta Polska*. By then it was obviously too late to withdraw or censor the issue of the journal, and the only official action was a formal reproach to the editors (*Schriftleitung*) of the *Zeitschrift* by the Ministry of the Interior, which had been informed about the offending map by the *Puste*.[444]

In addition to chance findings during their regular surveys of domestic and foreign publications, the *Publikationsstellen* also received maps from other institutions for evaluation. Government ministries specifically requested map evaluations from the *Publikationsstellen* to make decisions on whether to censor specific maps or to encourage their dissemination. For example, the Ministry of the Interior submitted two ethnographic maps by the Silesian division of the BDO to the *Puste* for examination in August 1939,[445] and the Foreign Office commissioned the *Publikationsstelle-Innsbruck* to prepare a report on a planned new edition of a map of the waterways in Central Europe in 1942.[446]

Even institutions which had no legal authority themselves to censor maps contacted the *Publikationsstellen*. They requested assistance in the elimination of maps which they deemed to be counter to German interests. The city archive in Elbing contacted Albert Brackmann in May 1938 and presented slides of a map in W. Eggers, *Harms Grosser Schulatlas*,[447] stating that the map supported Polish claims to West Prussia, because it showed that the corridor area had belonged to Poland in 1466. It asked Brackmann to initiate the establishment of clear guidelines and a regulatory process to prevent similar publications in the future.[448] Brackmann passed the offending map to the *Puste*, which in turn contacted the Ministry of the Interior, the Propaganda Ministry, the *Volksdeutsche Mittelstelle*, the Education Ministry, and the Foreign Office to get the map removed.[449] In its response to the city archive in Elbing, the *Puste* pointed out that it considered the eradication of such flaws its responsibility and encouraged the future submission of offending materials by the archive.[450] The archive willingly complied and sent several other maps.[451]

Unknown to the archive in Elbing, the issue of a clear regulatory process was very much on the minds of the *Puste* and other government-affiliated institutions in May 1938. Only a month earlier, Hitler had ordered the withdrawal and destruction of a map of German national territory and culture in the Nazi party publication *Der Schulungsbrief*.[452] Apparently the map had incurred Hitler's disapproval because it showed the German ethnic areas in South Tyrol, Italy. This claim to North Italian territory was politically undesirable since Hitler was going to meet with Mussolini and renounce German claims to South Tyrol. The cartographic transgression was considered serious enough to put the people responsible in jail and all institutions involved in the mapping of German national issues became extremely nervous; they were unsure of how to pursue their cartographic activities.[453]

However, Hitler's order to destroy the map came too late and some maps still went out. At least one copy reached Switzerland, where it caused a stir since it depicted most of Switzerland as German *Volksboden* (Grimm 1939: 13). To avoid future occurrences of this sort, the Propaganda Ministry issued a directive, which was also disseminated by the *Deutsche Kartographische Gesellschaft*.[454] It stated that South Tyrol could not be depicted as German on maps and that not even the name could be mentioned in publications. In addition, it urged publishers to exercise the utmost restraint in the propagation of maps dealing with German national issues, even if it resulted in economic losses.[455]

The decree was vehemently opposed by *völkisch* nationalists, who believed that this was treason to the Germans in South Tyrol, a view which was apparently even shared by some members of the Interior Ministry.[456] But the Propaganda Ministry kept the upper hand in the matter, since it had the backing of Hitler. South Tyrol, which was previously depicted as solidly German, was turned into Italian ethnic territory on ethnographic maps published in 1938.[457] The Propaganda Ministry, together with the Foreign Office, was also successful in removing all German *Volksboden* in the South Tyrol region from maps in the 1938 VDA exhibit *"Deutsches Volkstum im Ausland"* after it had already been approved by the *Volksdeutsche Mittelstelle* and the Interior Ministry.[458]

The facts that government organizations held opposing views and were overriding each other's decisions and that a party publication contradicted Hitler's political objectives underscore the lack of a well-defined cartographic regulation process. There were no guidelines on what should constitute German territory; it was not even clear who was in charge of issuing such guidelines. In response to this void and the associated confusion, different institutions began to take initiatives in the summer of 1938.

The *Puste* contacted the Ministry of the Interior and requested the

168

confiscation of three ethnographic wall maps which were being used in schools.[459] The suggestion was taken up by the Education Ministry and the Propaganda Ministry, which made sure that the maps were eliminated.[460] Likewise, the central office of the *Bund Deutscher Osten* made several successful appeals to the Propaganda Ministry to have offending maps removed.[461]

Apart from this reactive cartographic regulation, which was directed at removing existing maps, attempts were made to establish a preventative censorship process. The chosen fora were the meetings at the *Volksdeutsche Mittelstelle* (*Vomi*) on the production of a map of German national territory (see Chapter 8, "Coordination of Research"). In response to the *Schulungsbrief* incident, censorship issues were included in the agenda of the meetings. The participants were kept abreast of up-to-date information on censored maps and restrictions on the depiction of German territories,[462] and a decree was agreed upon which stipulated that every map dealing with political aspects of the German nation (*volkspolitische Karte*) had to be submitted to the *Vomi*.[463] A process was agreed upon, whereby maps were passed from the *Vomi* to the BDO for political evaluation, and then to the *Puste* for an examination of the scientific basis.[464]

Despite the decree, the regulation process never became formally established. In a lecture given in October 1938, at the second meeting of the *Deutsche Kartographische Gesellschaft*, it was stated that there was still no unified approach to regulation and that there was no cooperation among government institutions (Frenzel 1938: 375). In March 1939, half a year after the process had been initiated, map publishers still did not know to whom maps had to be submitted.[465]

A major problem in regulation was a lack of clear guidelines. Even the participants of the *Vomi* meeting were not sure who was in charge of issuing such rules, but there was hope that Alfred Rosenberg would clarify the situation in the near future.[466] Political developments in March 1939, that is, the creation of the protectorate Bohemia-Moravia and the incorporation of the Memel territory, prevented a decision and put the whole issue of depicting national territory in flux again. Cartographic regulation in National Socialism ended up being plagued by conflict, contradiction, and confusion.

Conflict was not only an outcome of the characteristic double structure of party and government offices in the Nazi state, which was intended to provide an internal control of power, but also the result of the varied nature of maps of national territory. Maps were used at home and abroad, for education, propaganda, and reference, by the party, the government, and private interest groups. Jurisdiction over maps often overlapped between different institutions and transgressions into the sphere of interest of an institution were vehemently opposed. For example, a regional office of the Propaganda Ministry (*Landesstelle*) recommended a map by Friedrich Lange for use in schools, which prompted the Education Ministry to disseminate a decree to education administrations in the German states (*Länder*) stressing that regional offices of the Propaganda Ministry had no authority to issue decrees about educational materials.[467]

Conflict even went so far that one regulatory institution checked up on the other. Even though the BDO was supposed to be part of the regulation process, the Ministry of the Interior requested that the *Puste* examine the shortcomings of two BDO maps.[468]

When the publication of cartographic depictions with incorrect political boundaries led to the confiscation of large numbers of maps in 1941, the Propaganda Ministry became very concerned that other ministries had taken part in this process and thereby impinged on its jurisdiction in cartographic censorship. To prevent future occurrences of this sort, it made it mandatory that from then on all cartographic representations had to be submitted to the Propaganda Ministry (*allgemeine Vorlage-pflicht*).[469] However, this did not prevent other institutions from continuing to be involved and from issuing contradictory regulations.

Contradictions in regulation were commonplace. For example, two school atlases by the Westermann Verlag, which were not approved for use in primary schools (*Volksschulen*) by the Education Ministry,[470] had been recommended by the *NS-Lehrerbund*, Bayreuth; one of them had even been called "extremely valuable."[471] There were contradictions not only concerning the appropriateness of individual maps or atlases, but also with regard to general principles of design, such as the choice of color for the territory of Germany and her allies, or with regard to the desirability of topics.

In 1941, the *Deutsches Zentralinstitut für Erziehung und Unterricht* ordered the Westermann Verlag to change the color it used to designate the Central Powers on its map *Westermanns Weltkriegskarte*. The publisher replied, rather perturbed, that the Propaganda Ministry and the German High Command had specifically requested the existing color scheme.[472] Disagreement about which color should be used to represent Germany also existed between the Ministry of the Interior, which preferred red, and the Propaganda Ministry, which had "other ideas on the subject."[473]

In January 1940, the *Reichspropagandaamt*, Berlin, pointed out in a secret press circular that it was not advisable to illustrate French war aspirations with maps of the Peace of Westphalia since this would draw attention to the fact that in this peace the Netherlands had been separated from Germany.[474] A few months later, in May 1940, the Nazi party publication *Der Schulungsbrief* (5. Folge, 1940: 78) used precisely such a map to illustrate the belligerent and anti-German character of the French.

These contradictions cannot simply be explained by competition among different organizations since contradicting messages were even issued by one and the same institution. Eight wall maps from the Westermann Verlag, which were prohibited in 1941 by the *Zentralinstitut für Erziehung und*

Unterricht, were listed as approved maps a few months later in the *Minister-ialblatt* 20/1942 for agricultural schools (*Landwirtschaftsschulen*).[475] Similarly, the Education Ministry prohibited a map showing the colonial development in the seventeenth and eighteenth centuries for upper-division schools (*höhere Schulen*), but at the same time made it a requirement for teacher colleges (*Lehrerbildungsanstalten*). The publisher was infuriated because the curricula of the two schools were virtually identical.[476]

Confusion about who was in charge and what was permissible was the consequence of the conflicts and contradictions in cartographic regulation and hindered cartographic production. When the Perthes Verlag wanted to produce a completely revised edition of the map by Haack-Rüdiger, *Das Deutschtum der Erde*, in March 1939, neither the publisher, nor the author, who was a participant of the *Vomi* meetings, were sure how to proceed.[477] The publisher knew that the Propaganda Ministry's authority was challenged by other institutions, and that nobody had the courage to issue definite guidelines after the *Schulungsbrief* incident regarding the depiction of South Tyrol. But he feared that contacting all institutions which claimed jurisdiction over German national issues — of which he knew at least eight existed — would delay publication too long.[478] As a solution he opted to issue a slightly modified edition disguised as a simple reprint. To avoid possible difficulties he made one institution his "protector," that is, he got the unofficial backing and approval of one institution he believed had most to say about wall maps, the *Parteiamtliche Prüfungskommission*.[479]

Confusion was also caused by tardiness. In 1937, the Education Ministry began to prepare the substitution of the existing atlases for primary schools with new atlases which were to correspond more closely to the educational principles and ideology of the Nazi party. Although the Education Ministry held conferences and published commentaries, the details of this reform were only known to a small circle of select people.[480] On 25 February 1938 and on 18 February 1939, the ministry issued decrees which stipulated that drafts for new primary-school atlases had to be submitted for evaluation.[481] Two atlases submitted by the Westermann Verlag were not approved, with the argument that they did not correspond to the decree of 25 February 1938. With the rejection notice, the publisher also received for the first time the general guidelines on the basis of which he was supposed to have drafted the new atlases, but which the ministry had failed to append to the decrees requesting submission.[482]

As a result of conflicts, contradictions, and confusion, enforcement of a unified message in maps during the Nazi period proved to be difficult. For example, because the approved drafts for the new primary-school atlases were not ready to be disseminated, the use of existing primary-school atlases was tolerated, which enabled the Westermann Verlag to tacitly distribute its unapproved primary-school atlases.[483] The lack of enforcement regarding the new primary-school atlases was not a unique case. In February 1940, the secret police reported that many Czech-language schools in Vienna still used maps which showed the former Czechoslovak Republic and also historical maps which outlined far-reaching claims to German territory.[484]

There were still problems in the summer of 1943. Hungary and Bulgaria officially complained about the incorrect depiction of German political boundaries in the map "Der deutsche Osten" in the semi-official journal *Das Reich* from 11 July 1943. This prompted the Foreign Office, the Ministry of the Interior, the *Reichsamt für Landesaufnahme*, the Propaganda Ministry, the German High Command, and the Ministry of the East to initiate a pre-publication censorship of maps in magazines and newspapers.[485] A resolution of 26 August 1943 stated that, with immediate effect, all cartographic depictions with political, historical, and economic topics in newspapers and magazines had to be submitted to the Propaganda Ministry and the *Reichsamt für Landesaufnahme* if they covered the German Empire, the *Generalgouvernement*, and civil administration areas (*Zivilverwaltungsgebiete*). Maps which depicted areas abroad had to be submitted also to the Foreign Office, and maps of the occupied territories in the East to the Ministry of the East. Publishers were advised to speed up the process by presenting the maps to the four institutions via messengers.[486] There are no records about further regulatory problems, but this should not be surprising. Faced with such an involved process and high costs for using messengers, most publishers of newspapers and magazines under war-time constraints probably opted not to use maps at all.

INNOVATIONS

Despite the reluctance of the Nazis to take over complete control of propaganda mapping, their advent to power brought about changes in suggestive cartography which went beyond a mere collaboration with existing institutions. However, these were not initiated by National Socialists, but were the result of the same type of voluntary submission to the political needs of the Nazi state that prompted geographers to adopt racial terminology and concepts (see Chapter 8).

Previous suggestive cartographic activities – such as the regular VDA/ DAI map service for newspapers and magazines, the use of suggestive maps in schools (e.g. Sämer 1935), and the publication of inexpensive suggestive atlases (e.g. Pudelko and Ziegfeld 1938; Lange 1937) – were continued with a renewed sense of pride in being German.[487] Accompanying texts for atlases or the wording of captions were more and more assertive in their advocacy of German superiority. One academic felt that this new national self-esteem was not flaunted enough. Kleo Pleyer, the geographer in charge of the *Volkswissenschaftlicher Arbeitskreis* of the VDA, urged other scholars to make sure that maps of the entire German settlement area, such as Friedrich Lange's *Sprachenkarte von Mitteleuropa*, were prominently displayed in university buildings.[488]

The general attempt to boost public confidence and involvement in German national issues after the Nazi seizure of power is further underlined by a new VDA project. The organization started a series of map quizzes on German issues (*volksdeutsche Schulungsaufgaben*) in the January 1935 issue of the *Rolandblätter*, its monthly magazine for young people. The quiz gave the outline of foreign territories with German settlements, such as Alsace-Lorraine or Czechoslovakia, and asked the reader to identify the state or region depicted as well as to answer a number of questions on Germandom. The next issue of the periodical contained the correct replies together with a suggestive map.[489]

There were also new forms of cartographic propaganda which reflected the preference of the Nazi regime for emotive images and public displays.[490] Maps gave increased attention to allegorical representations designed to engender an emotional response. For example, the threat from Czechoslovakia was indicated by a fist (see figure 9.4), and the dissection of the German settlement area by political boundaries was illustrated with a knife (see figure 9.5).[491] These depictions appealed to the primal instinct, to the struggle for survival. Like an actual human body, Germany was in danger of being mutilated or killed by her enemies: punched, strangled, and cut. The message was unambiguous: Germany had no choice but to counterattack before it was too late.

Maps in displays became more prominent in the Nazi period. The *Deutsches Ausland-Institut* in Stuttgart, which attempted to become the main institution supplying materials for exhibitions and political education (*Schulung*) after 1933 (Ritter 1976: 71), created unique exposition pieces which used suggestive maps: German national "shrines" (*volksdeutsche Schreine*). They focused on different regions outside the state boundaries which had a presence of Germans, such as Bohemia. The displays were shipped in three handy boxes which could be assembled to look like a house altar with three wings. The back panel contained photos of landscapes and suggestive maps of the settlement areas of Germans abroad. The boxes in front of these panels could be folded out to form a horizontal surface for relief models of farmsteads and craft artifacts.[492]

Giant public displays, a favorite of totalitarian systems, also found cartographic expression. In contrast to the permanent character of the Italian marble wall maps on the Forum in Rome, Nazi maps were more temporary. Therefore, public map displays in Nazi Germany were well suited to adapt their themes to changes in propagandistic emphasis. Early maps expressed specific territorial demands, while later versions stressed military threats and attempted to get public support for war.

In 1933, miners from Rüdersdorf made a giant map out of brick and

Die Oſtfront auslanddeutſcher Städte

Figure 9.4 Allegorical Map I: "The eastern front of German cities outside the German state territory" (Lange 1937: 32).

chalk which demanded "peace and equality" (*Frieden und Gleichberechtigung*) in large letters. Using a dark background surface, it depicted the territory of the German state in solid white and the pre-Versailles boundaries with clearly visible white lines. The territorial demand this entailed was clear: a complete revision of Versailles.[493]

During a major public event in Berlin, the "*Fest der deutschen Schule*," which

Figure 9.5 Allegorical Map II: "German people, you are dissected by boundaries"
(Oerdingen and Stein 1934: cover).

was part of the "*Tag des deutschen Volkstums*" (10 September 1933), the
VDA organized a cartographic play intended "to show the multiformity
and fragmentation of the German *Volkstum* in the world," and to "forge"
a sense of unity. The play was performed in the German Stadium in

175

Berlin-Grunewald and had giant proportions; it involved 6,000 participants (students from Berlin and members of the *Trachtenarbeitsgemeinschaften Deutscher Landsmannschaften*). The main theme was the separation of Germans into different territories and the ensuing need of the German state to reach out and unite them all. The sequence of the individual scenes reveals the revisionist character of the play:

1 Representatives of different German regions and cities in traditional costumes and colors position themselves on the stage within the political boundaries of Germany, which are displayed in black–white–red.
2 Representatives of the different regions ceded in the Treaty of Versailles appear and form the outline of their respective territories. In the Saar region, French soldiers surround the German representatives to show the occupation. In Danzig, the Germans are crowded by foreigners
3 Representatives of other German areas abroad appear.
4 "Cry for help to the German Reich." The Reich responds by hoisting the German flag. The actors notice this and shout: "*Heil.*" The pledge to the flag is given, the German political boundaries disappear, and representatives in Germany and abroad join hands.
5 The swastika flag is formed by all actors.
6 The play ends with a torchlight procession.[494]

By the late 1930s, the focus of public map displays had shifted from specific territorial demands to threats from abroad. A map at the 1938 Nazi party convention exhibit used red light bulbs to show the lines of a possible Bolshevik attack from the air via Czechoslovakia and by ship via the Baltic and Black Seas.[495] Once Nazi Germany had started the war and gained victories, public displays stressed German military achievements rather than vulnerability.

The German military produced films which employed cartographic trick sequences to portray the far-reaching advances of the German armies, such as *Der Feldzug in Polen* (1939) and *Sieg im Westen* (1941).[496] A contemporary witness, Hans Speier (1941: 326) described the "moving maps," as he called them: "On the screen, movements resemble attacks of bacilli upon live matter, or the action of phagocytic cells observed under the microscope. Like these processes, the advance of the Germans seems to be removed from ordinary sense impressions." Speier explained why the forward thrusts of troops in the maps were shown as a sequence of abrupt movements: he believed that the intention of the German propagandists was to convey not natural and smooth processes, but a sense of "controlled mechanical effort" (p. 326).

The trick sequences for both movies were made by professional trick film ateliers and no credit was given to suggestive cartographers. Yet, some of the maps in the movies seem to have been inspired by existing suggestive maps, such as a map trick in the movie *Sieg im Westen* which

176

depicted the military threat from Czechoslovakia as arrows moving toward Berlin, Munich, and Breslau, an apparent adaptation of the map by Schumacher (1934) in figure 9.3.[497]

The *Gaupropagandaamt* in Oberdonau made a unique contribution. It erected a large (ca. 4 × 5 m) billboard with a map of Great Britain in a central square in Linz in the fall of 1940 to show German military successes. The map showed Great Britain, Ireland, the southern tip of Norway, and parts of the west coast of the European mainland. A line of symbols surrounding Great Britain and Ireland gave the impression that the British Isles were completely isolated by a sea blockade. Another set of symbols represented exploding ships, burning factories, and artillery bases. They were constructed as colored wooden cut-outs and placed in such an abundance on the southern British coast and around London that the map left little doubt of an impending German victory. The map, which took up most of the billboard, was supplemented with a legend, a caption, and two small political cartoons in the bottom left and bottom right corners. The *Gaupropagandaamt* claimed that more than 100 people assembled before it on Sundays, that the board was the talk of the town, and that Rudolf Hess had expressed his appreciation of it during a visit to Linz. The *Gaupropagandaamt* planned to erect similar billboards in other towns, but there was no further correspondence in the files which would indicate that the plan was carried out.[498]

Propaganda mapping in Germany continued until the end of the Nazi regime, but there was a definite decline in importance during the war period. The shift in emphasis in maps of national territory toward government application and secrecy by the late 1930s, the extension of German control over even the most extreme German claims to national territory by early 1940, and the eventual prohibition of maps dealing with German national issues in 1942, deprived the suggestive cartography network of most of its influence. In addition, the war meant that the military, which obviously had its own agenda, became involved more and more (Welch 1993: 90). Specific territorial claims, for which maps are the ideal means of communication, did not suit military requirements very well. Propaganda needed to be flexible to adjust to changing military conditions and, especially after the loss at Stalingrad, propaganda retreated "more and more into the mythical and irrational elements of National Socialist doctrine" (Welch 1983: 6), which were difficult – if not impossible – to represent cartographically.

CONCLUSION

The feeling of belonging to a nation is not based on primordial kinship, it is socially constructed around three main elements: a common past, a common destiny, and a common territory. For Germans, a national identity emerged in the early 1800s, but it took more than a century before it became important to define the precise limits of their national territory. The stimulus came from the outside: at the end of the First World War, the Allied and Associated Powers declared that the principle of national self-determination was to become the basis of the peace treaties. This meant that the political boundaries of Europe were to be brought into agreement with the distribution of national groups. Maps played a crucial role in this process: they were used to delimit the new national boundaries and to present nationalist claims to the victorious powers. Only maps offered a clear territorial image and an effective means of communicating territorial limits to others.

The Germans were very slow to recognize the need to define the extent of their national territory and the value of maps in this process. Before the German government publicly embraced the principle of self-determination as a basis for peace, the precise delimitation of the extent of the German nation was not a pressing issue. From the beginnings of German nationalism, attempts to create a German national state had striven to unify the multitude of existing German principalities rather than to define new boundaries; even though this meant the inclusion of foreign nationalities.

The German government and the German population erroneously believed that the acceptance of Wilson's Fourteen Points in the negotiations leading to the armistice would result in a favorable peace. It was not until the territorial cessions stipulated in the Treaty of Versailles were presented to the Germans that they became painfully aware of the need to address the issue of national territorial limits. However, the German government was not very interested in using maps. This prompted German scholars, who were more aware of the need to map German national territory, to take initiatives. Realizing that they had neglected research on German national territory, German geographers, such as Albrecht Penck, had already started mapping projects around the time of the armistice. But their efforts came too late and were not appreciated by the government.

Further sensitized by the territorial losses in the Treaty of Versailles, German scholars intensified their efforts. Initially, a few scholars embarked on individual projects: they exposed flaws in foreign ethnographic maps and prepared new maps which supported claims to Germany's pre-war territory. In the absence of a strong government interest, they cooperated with *völkisch* groups, such as the *Deutscher Schutzbund*, and orchestrated a comprehensive attack designed to maximize claims to German national territory. They developed not only more accurate cartographic methods, but also new spatial concepts of national territory: the geo-organic concept, which argued for the indivisibility of the "organism" of the pre-war German Empire; the negative definition of national territory, which stressed that the new boundaries made Germany vulnerable to attack; and the *Volks- und Kulturboden* concept, which pointed out manifestations of German culture in the landscape. By the mid-1920s, their success in outlining claims to more territory, combined with German economic prosperity, refurbished German national self-esteem from its all-time low after the defeat. This fostered a belief in the cultural superiority of Germans and sparked the production of maps of Germany's far-reaching cultural heritage in the East.

All new maps – whether they depicted spatial concepts of national territory or the influence of German culture – reached a wide audience because the combined efforts of scholars and *völkisch* nationalists had also focused on effective representation and the control of dissemination. By the end of the 1920s, these scholars and nationalists had created a network of institutions engaged in the production and distribution of suggestive maps and managed to establish a system of coercion which ensured a unified nationalist message. Surprisingly, the government stayed in the background during the entire period, even in regulation. It gave some financial support, but only became active in response to outside suggestions or foreign activities.

When the Nazis came to power in 1933, they encountered a situation which proved remarkably useful for their endeavors. The German public had been exposed to numerous maps which showed that German national territory included Alsace-Lorraine, Austria, the Memel territory, as well as parts of Poland and Czechoslovakia. Some of the maps even laid claim to a region of German cultural heritage which reached as far east as the Ukraine and essentially matched Hitler's goals for *Lebensraum* (see figure 10.1).[499] In addition, the German public had been presented with numerous maps which showed that the present German boundaries made the German nation vulnerable to attack. This gave the clear message that military build-up and expansion were necessary to obtain a more easily defensible territory, which was a central demand of National Socialism.

179

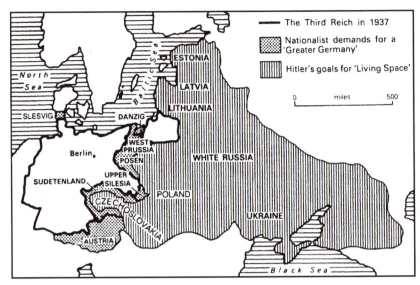

'Lebensraum'; plans and projects for German expansion in the East.

Figure 10.1 Hitler's goals for *Lebensraum* (Wright and Stafford 1938: 14).

The Nazis could also count on collaboration by individuals and organizations involved in issues of national territory. Experts on suggestive mapping offered their help and a sophisticated propaganda mapping network with an effective system of coercion was in place. For practical applications, such as planning of annexations, conquests, or administration, research that was orchestrated by the Leipzig *Stiftung für deutsche Volks- und Kulturbodenforschung* in the Weimar Republic had yielded a wealth of ethnographic data, more accurate methods of cartographic representation, and a legion of specialists.

The Nazis made use of what was put at their disposal. Initially, they took over the new spatial definitions of national territory, which had been developed in the Weimar Republic, without alterations. The geo-organic definition was used to show the dismal state of Germany at the time of the Nazi ascendancy to power, while the negative definition gave strong support to rearmament and the lifting of military restrictions. The clearly defined territory of the *Volks- und Kulturboden* was too limiting for the Nazis. It was tolerated only when it was combined with depictions of the German cultural heritage in the East, which gave some flexibility.

In the late 1930s, the Nazis saw the need to modify the existing spatial concepts to ensure maximum compliance with National Socialist goals. When territorial gains were made which went beyond the pre-First World War boundaries of the German Empire, such as the Sudetenland, the geo-organic definition faded into obscurity. When Germany rearmed success-fully, the negative definition abandoned its focus on general strategic shortcomings, and the threat to German territory was "personalized" with depictions of the hostile intentions of specific countries. This helped prepare the German people to go to war by creating foe-images. When the Nazis intensified the execution of their racial program around the time of the *Reichskristallnacht*, the *Volks- und Kulturboden* was "cleansed" of geo-graphic or cultural elements. The idea of a German *Kulturboden* was discarded because it allowed for the transformation of other ethnic groups into "cultural" Germans, and the definition of the *Volksboden* was given an explicitly racial slant by introducing anti-Semitic principles. This shift from spatial definitions of national territory back to person-based or social definitions was mirrored in the emergence of new depictions stressing pre-historic German territory. However, despite the central role of racism in Nazi ideology, explicitly racial definitions of national territory did not become prominent because they severely restricted territorial claims.

For the propagation of these different concepts, the Nazis could make use of the well-functioning network for propaganda mapping and coercion which had been established in the Weimar Republic. They did not have to expend time or effort to develop propaganda mapping on their own. The cooperation with existing institutions fitted their needs so well that they did not even elect to centralize or standardize production and dissemination of maps; despite important theoretical advances and the voluntary assistance of accomplished cartographers which would have allowed them to do this easily. The passive attitude of the Nazis also affected regulation and censorship, which were mired by conflict, contradiction, and confusion. They only restricted the publication of maps when the message was in direct opposition to Nazi policy. Their main contributions to propaganda mapping were the introduction of emotive allegorical symbols and the use of maps in public displays.

There are two main reasons for the lack of Nazi involvement in propa-ganda mapping. First, the maps were used for a variety of purposes and fell under the jurisdiction of many different authorities, which made their organization and control very difficult. Second, *völkisch* nationalists engaged in the existing structures could be left in charge because they shared many of the ideas of National Socialism and in many cases were willing to adapt if there was a divergence. An example would be Karl C. von Loesch's sudden interest in racial issues after the Nazi advent to power. This is not to say that there were no differences between *völkisch* nation-alists and National Socialists. *Völkisch* nationalists had a very clear agenda, namely the incorporation of all Germans in one state, and they pursued this even if it was politically inopportune. By contrast, the political agenda of the Nazis was absolute power and they were willing to sacrifice anything and anybody to achieve it. When Hitler renounced German claims to South

181

Tyrol and ordered the depiction of this region as Italian territory, *völkisch* nationalists, such as Friedrich Metz, found this to be treason against the German majority living in the area. Similarly, *völkisch* nationalists had an historical focus. They were concerned about the roots of German settlement and where Germans had "imprinted" their superior culture on the landscape, while the Nazis had an activist bent and were interested in the lands on which Germans could "imprint" their superior culture. The main conviction they shared was that German culture was superior.

The opportunistic attitude of National Socialists to the mapping of national territory is also revealed in the way they exploited maps for their changing needs. When Nazi Germany prepared for conquest in the late 1930s, it not only encouraged modifications of the different definitions of national territory, but also supported the practical application of maps for the execution of its expansionist policies. This meant a renewed emphasis on accuracy in ethnographic maps. Detailed and reliable maps were needed to delimit the German territories the Nazis demanded from Czechoslovakia during the Munich Conference. Knowledge of the precise distribution of national groups in Eastern Europe provided the basis for taking control of these vast areas; it allowed the Nazis to instigate civil wars which played off different national groups against each other. Finally, reliable ethnographic maps proved indispensable for the Germanization of occupied territories. The research supported by organizations during the Weimar Republic, such as the Leipzig *Stiftung*, served the Nazis well. However, there is a significant difference between the Nazi concern for accuracy and the corresponding politicization in the 1920s. The Weimar version stressed publication to bring about a public consensus. The Nazi version was classified to ensure unopposed execution of the policies for which the maps were designed.

Surprisingly, the Nazis did not prohibit the dissemination of maps of German national territory until the summer of 1942, even though they had achieved territorial gains far beyond German national territory much earlier. The Nazis only discarded the demand for German self-determination and pursued the development of alternative concepts when they had consolidated their control over the occupied territories and believed that the war would end soon.

In general, because respected scientists and well-known *völkisch* thinkers had already given credibility to far-reaching claims in the Weimar Republic, their continued use under the Nazis – albeit in modified versions – was very effective. Although there was a wide variety of claims, they all demanded at least the Sudetenland, the Memel territory, Austria, and parts of France and Poland as German national territory. This reinforced rather than diluted the argument for expansion in the name of national self-determination. At the same time, variations in the extent of further-reaching claims left things flexible and avoided setting a definite limit for conquests.[500]

The mapping of German national territory presented in this book has important implications for the current debate about the place of the Nazi period in German history. This debate was brought into the center of political discourse by the attempt of neo-conservative historians to show that the Holocaust was not unique and by the need to redefine German identity after unification. It is clear that the German public was not lured into a war of conquest by skillful Nazi propagandists, but was already prepared for expansion by the combined efforts of geographers and *völkisch* nationalists in the Weimar Republic. Historical continuity is apparent not only in expansionist goals, but also in the voluntary and even eager collaboration by experts and organizations. The Nazi state was not an aberration in an otherwise "normal" German history and Germans were not simply victims of a powerful regime.

Outside the larger context of German national identity, the findings necessitate a re-evaluation of existing views on propaganda maps. The majority of examples of effective propaganda maps that are presented in the literature are products of totalitarian countries or of powerful state institutions, such as the CIA or the military. This gives the impression that the use of maps in political propaganda requires the direct involvement of the state. The examination of German propaganda mapping shows that a private network is completely sufficient to propagate a unified message. The only prerequisites seem to be that the network is united by a clear political goal – in the German case the revision of the Versailles Treaty – and that an organization or an individual takes a leading role in the beginning. In addition, contrary to the view presented in the literature, National Socialism, and specifically the school of *Geopolitik*, cannot be considered the epitome of cartographic propaganda. *Geopolitik* only played an adjunct role in the development of suggestive mapping, and the involvement of National Socialism was negligible. German propaganda cartography was pioneered long before the Nazis came to power and by a variety of groups and individuals, with the most decisive influence coming from *völkisch* groups.

The involvement of scholars in the creation of a consensus about German territorial demands is also instructive. *Völkisch* groups would not have been as effective in securing funds for projects, in enforcing a unified message, and in disseminating their maps without help from respected scholars, such as Albrecht Penck and Wilhelm Volz, or from professional organizations, such as the *Deutscher Geographentag*. This entanglement of scholarship in political persuasion was not simply the result of "biased" work; German academics still played by the rules. Rather, it was an outgrowth of the general political nature of academic work; academic work is political because it forms the basis for political and ethical decisions.

183

Scholars cannot hide behind the idea of value-free and objective research to avoid responsibility – such as Troll's (1947) attempt to give absolution to German geography from Nazi involvement – scholars always should be aware that their work can be used as a powerful tool for political persuasion. Reference to "impartial" research gives political actions a cloak of respectability. This issue is particularly relevant in the context of the *Historikerstreit*. The dispute among historians in the late 1980s about the uniqueness of the Holocaust and Germany's war guilt was not just an academic affair. Without support from well-known historians and intellectuals, the New Right in Germany would not have been able to broaden its base of supporters well into the societal mainstream. Awareness of this political nature of scholarly work has to be extended beyond the academic profession itself to the general public, especially in view of the special respect professors enjoy in Germany.

Intimately connected with the political nature of scholarly work is the issue of "just" national boundaries. There has been a phenomenal increase in demands by national groups to be granted sovereignty over their rightful territory. But as with the dispute over the Polish or German character of the corridor, the delimitation of national boundaries is always arbitrary, no matter how scholarly the procedure. On the one hand, there is no general agreement on definitions of national territory. On the other hand, national groups are not neatly differentiated from each other. They always spatially intermix in the areas where they meet. Our perceptions that clear national boundaries are possible are influenced by the most commonly used cartographic method to depict national groups: areal shading. In this method, areal units, such as administrative districts, are colored in the hue of the group which has a majority in the area. This completely obfuscates the presence of minorities in the same area. An alternative method, such as Albrecht Penck's graduated dot map, shows the intermixed distribution which exists in reality. It also shows that it is impossible to draw a "just" boundary through this maze of interspersed settlements.

The difficulties in delimiting national territories expose national self-determination as a dangerous political principle. The demand to extend political sovereignty over one's "rightful" national territory cannot be granted to one group without inciting opposition from another. This calls into question the permanence of territorial divisions along ethnic lines which are attempted in Bosnia. Ultimately, we will have to re-evaluate the usefulness of the idea of territorial sovereignty and work towards dividing political power into different territorial layers.

Since the collapse of the Third Reich in 1945, only a few maps of German national territory have appeared. In the German Democratic Republic, the press and other publishing activities were severely censored and maps staking national territorial claims would have been incompatible with the communist ideology of the government. In the Federal Republic of Germany, there was a similar paucity of maps of national territory until the late 1980s. The only cartographic depictions of national territorial issues which appeared earlier were intimately connected to Germany's defeat and occupation status. They either portrayed the fragmentation of Germany into West, East, and areas under foreign administration – accompanied by headlines such as "Divided in Three? Never!" – or they laid claim to the territory of "Germany within the borders of 31 December 1937." This date has special significance because it was used in the London Protocol of 12 September 1944 to define the limits of the occupation of Germany and has also been incorporated into the Basic Law of the Federal Republic of Germany (article 116) to define German citizenship.[501] Before German unification in 1990, nationalist groups such as the *Vertriebenenverbände*[502] insisted that the 31 December 1937 borders were the only legal basis for the boundaries of the German state. They demanded that all German maps, especially those used in education, should clearly indicate them.[503]

In the late 1980s, a newly founded right-wing party, *Die Republikaner*, initiated a small renaissance of maps of German national territory with the dissemination of three maps: a map of the German boundaries according to international law, a map of German settlements prior to the outbreak of the Second World War, and a map of Germany's former eastern territories.[504] These maps mirrored the demands and designs of maps made by *völkisch* groups in the Weimar Republic.

The map of Germany's legal international boundaries made reference to the right of national self-determination. It demanded Germany's pre-First World War boundaries combined with Austria, the ethnic German territories of South Tyrol (Alto Adige), Travis, and the so-called Sudetenland, which was given to Nazi Germany at the 1938 Munich Conference. The map of German settlement areas before the outbreak of the Second World War goes even further. It is essentially a reprint of the Weimar period map "*Deutsches Grenzland in Not*" (see figure 7.1).[505] The third map is a variation of the geo-organic concept and shows agricultural losses and the number of Germans in the eastern parts of pre-First World War Germany and the Sudetenland (see figure 10.2).

The electoral success of the *Republikaner* party has faded since German unification in 1990. But it is unlikely that such territorial demands will disappear from the political scene. A major factor is the national identity crisis which has befallen Germany since unification. As long as Germany was divided in two, the German national discourse was effectively severed from its historical roots. The East created an identity around the idea of communist brotherhood and the West around economic prosperity, a D-Mark nationalism. The unification of Germany not only revealed the vast cultural, political, and economic gap between East and West, it also reopened the German question. Where can the German nation find its roots? In Bismarck's German Empire, which carries with it the stigma of

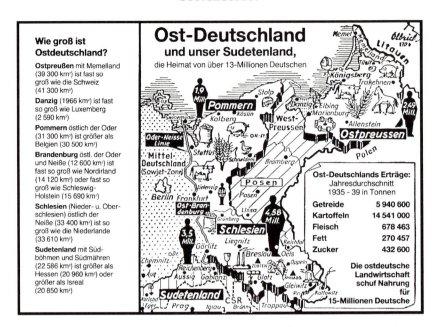

Figure 10.2 Contemporary map of German national territory: "East Germany and our Sudetenland, home to more than 13 million Germans" (map disseminated by the German party *Die Republikaner*, no author, ca. 1989).
The text on the left side gives the areal size for the different regions depicted on the map and makes comparisons with other territorial units. For example, the Sudetenland is compared with Hesse and Israel. The text insert on the right side lists the annual average for harvests in tons between 1935 and 1939 for grain, potatoes, meat, fat, and sugar and states that eastern German agriculture provided food for 15 million Germans.

being the "Hun," the stigma of Prussian militarism and imperialism? Or in the ill-fated Weimar Republic, which continuously struggled for public acceptance? Neither one of these choices can get around dealing with the legacy of the Nazi period that followed them. We have not seen the last of the *Historikerstreit* and probably not the last map of the territory of the "Greater" German nation.

NOTES

INTRODUCTION

1 Cited after Demandt (1990: 15). All translations are by the author unless noted otherwise.
2 Cited after Rosenberg (1939: 15).
3 A similar demand is expressed in their 1990 party program: "Our demand: The reinstitution of Germany . . . The German question remains open until a peace treaty!" (Parteiprogramm 1990, *Die Republikaner*). The map identified the following territories as comprising the internationally legal boundaries of Germany on the basis of the right to self-determination (*Die völkerrechtlich gültigen Grenzen des Deutschen Reiches auf Grund des Selbstbestimmungsrechts der Völker*): the German Empire in 1914 (however, without the territories ceded to Denmark in the Treaty of Versailles); the Sudetenland; Austria's current territory with the additional regions of South Tyrol, Travis, and around Ödenburg (denoted as "areas annexed after 1919"). On the right side of the map image, which measured 68.5 × 61 cm, was a lengthy text justifying these territorial claims. The justification made reference to various international treaties and conventions.
4 For an overview of the *Historikerstreit*, see Baldwin (1990), Jarausch (1988), and Welch (1993: 129–35).
5 The strong currency of this scholarly reworking of German history, which spurred the "historians' dispute" (*Historikerstreit*) in the late 1980s, can be seen in the recent scandal involving Manfred Kehrig, the head of the Federal Military Archive in Freiburg. Kehrig wrote a preface to a revisionist work by Joachim Hoffmann in which he openly endorsed the arguments presented in the book. The book in question is Joachim Hoffmann, *Stalins Vernichtungskrieg 1941–1945*. Munich: Verlag für Wehrwissenschaften, 1995.
6 Although the term "*völkisch*" is commonly used to define the radical and anti-Semitic *ideology* that was a precursor to National Socialism (e.g. Mosse 1981), the *völkisch movement* was by no means a uniform entity. Not all of its members embraced the explicitly racial aspects of the ideology. Its organizations, such as the *Volksbund für das Deutschtum im Ausland*, even included members who were Jewish. An example is the publisher Franz Ullstein (see Chapter 7, "A Concern for Boundaries and Place Names"). The movement was held together by the belief in the superiority of the German people (*Volk*) – which for some entailed a cultural superiority, for others a racial superiority – and by the goal to preserve the unity of the German *Volk*, which meant the creation of a state that included the Greater German nation.
7 For a survey of the literature which comes to this conclusion see Welch (1993: 1–4) and Kershaw (1989: 1–17). See also Hamerow (1983).

1 NATIONALISM, TERRITORY, MAPS AND PROPAGANDA

8 Lord Acton (1907). Nationality. In *History of Freedom and Other Essays*. Salem, N.H.: Ayer Co. Cited after reprint in Dahbour and Ishay (1995: 113).

9 Institute for Propaganda Analysis (1937). How To Detect Propaganda. *Propaganda Analysis* vol. 1, no. 2. Cited after reprint in Jackall (1995: 217).

10 There is also no universal agreement about the concept of self-determination. Knight (1984: 168) defines it most generally as "the right of a group with a distinctive politico-territorial identity to determine its own destiny."

11 Only in this case, when the territory of the nation is congruous with the territory over which political sovereignty is exercised, can we speak of a "nation-state." The term "nation-state" is widely abused to connote states.

12 For a discussion of the spatial aspects of nations and states for the case of Canada, see Kaplan (1994).

13 For a contextual analysis of the historical development of mapmaking, see Edney (1993).

14 For a treatment of the theoretical context of objectivity in research, see Natter, Schatzki and Jones (1995).

15 For a discussion of the role of literature in German nationalism, see Seeba 1994; for the role of political symbols, see Mosse (1975).

16 See also Jackall (1995: 2), who argues that propaganda is "the product of intellectual work that itself is highly organized" and whose success depends on how well it "rechannels specific existing sentiments."

17 The title of the map is: Karl Bernhardi, *Sprachkarte von Deutschland. Als Versuch entworfen und erläutert*. Kassel: J. J. Bohne 1844 (Cited after Weidenfeller 1976: 409; and Fittbogen 1927: 74).

18 Langewiesche (1992b: 351–2). The uncertainty about the territorial extent of the German nation can be gleaned from an excerpt of a speech by a deputy of the parliament which is cited in Langewiesche (1992a: 64–5).

19 Karl Bernhardi, *Sprachenkarte von Deutschland, 2. Auflage, unter Mitwirkung des Verfassers besorgt und vervollständigt von Wilhelm Stricker*. Kassel 1849 (cited after Weidenfeller 1976: 409, fn. 147 and Fittbogen 1927: 74).

20 Cited after Weidenfeller (1976: 63). Emphasis added.

21 H. Kiepert, *Völker- und Sprachenkarte von Deutschland*, 1867 (cited after Isbert 1937a: 492).

22 Breuilly (1992: 12) argues that many nationalists opposed what Bismarck had done: the Confederation was destroyed and Austria was excluded.

23 The fact that many German-speaking Alsatians identified with France was not incompatible with German views on national self-determination. Heinrich von Treitschke argued that they were nevertheless German and that they had to be given back "their own self against their will" (Holborn 1969: 221–2).

24 On the *Alldeutscher Verband* (founded in 1890), see Wertheimer (1924) and Schödl (1978).

25 On the involvement of geography in German expansionism, see Sandner (1994: 82–7).

26 The small following of the ethnic national movement can be seen in Breuilly's statement that "by 1914 few German speakers in the German Empire thought of Germany as anything other than the territory of that Empire" (1992: 14).

27 The ideology of the *Alldeutscher Verband* was by no means well defined or uniform; it was represented by the views of its senior members rather than by its official program (Schödl 1978: 60). It also included aspirations for

expansion into overseas colonies. A common belief was that Germans who emigrated would quickly lose their identity abroad through intermixing with other cultures. The only solution to this perceived danger was the foundation of German colonies. There was considerable overlap in membership and cooperation between the *Alldeutscher Verband* and other nationalist organizations, such as the *Flottenverein* (see Eley 1980: 156–9).

28 On *großdeutsch* tradition and German expansionism, see also Schieder (1992: 58–63) and Weidenfeller (1976: 9–11). An example of the extreme *völkisch* position was Ernst Hasse, who was a member of the German parliament; see Schödl (1978: 56–60) for a discussion of his views.

29 This was admitted by Wilhelm Volz, who even extended blame to himself (W. Volz 1942: 724).

30 See also Schultz (1995), who cites the writings of the German geographer A. Kirchhoff to show that the notion to equate the German nation with the newly formed German Empire became influential after 1871.

31 For a general treatment of these organizations, see Possekel (1986) and Weidenfeller (1976).

32 Hermann Kiepert, *Übersichtskarte der Verbreitung der Deutschen in Europa, für den Deutschen Schulverein zusammengestellt*, 1:3,000,000. Berlin: Dietrich Reimer, 1887. Hermann Nabert and Richard Böckh, *Verbreitung der Deutschen in Europa. Im Auftrage des Deutschen Schulvereins*, 1:925,000. Glogau: Flemming, 1891. (Cited after Fittbogen 1927: 76.)

33 Paul Langhans, *Deutscher Kolonialatlas*. Gotha: Justus Perthes, 1897 (cited after Kirchhoff 1898: 13); *Alldeutscher Atlas*. Gotha: Justus Perthes, 1900 (cited after Fittbogen 1927: 76). Langhans also edited the journal *Deutsche Erde* (founded in 1902), which was dedicated to the ethnography of the German people and their culture and which was accompanied by color maps (see advertisement in Langhans 1905). For a list of maps in the *Deutsche Erde*, see Pessler (1931: 67). A list of Langhans's maps on the distribution of Germans is included in Langhans (1905).

34 Langhans was a member of the *Alldeutscher Verband* and the *Deutscher Schulverein* (later called VDA) (Eley 1980: 381). He also founded the Gotha branch of the small *völkisch* organization *Deutschbund* (Köhler 1987: 168).

35 A third edition of the atlas appeared in 1905 (Langhans 1905).

36 This double purpose is mentioned less explicitly by Eckert in a comment on Langhans's maps (Eckert 1925: 463).

37 Schieder (1992: 58–60). See also Weidenfeller (1976) and Schödl (1978) for the Austrian origins of the *Deutscher Schulverein* and the *Alldeutscher Verband*.

38 The geographer Wilhelm Volz, who collaborated closely with *völkisch* nationalists after the war, stated in his academic memoirs that before the war nobody in Germany expected that claims to the existing territory of the German state would ever have to be defended (W. Volz 1942: 723–4).

39 Penck was a member of the VDA (Faber 1982: 398).

40 It is not surprising that Penck and Sieger were among the first to take initiatives (see Chapter 2, "German Initiatives").

41 For a discussion of Germany's imperialist ambitions, see W. Smith (1986). German geographers provided intellectual and scientific backing for expansion into overseas colonies and *Mitteleuropa*; see Faber (1982), Böge (1987), Meyer (1946 and 1955), Sandner and Roessler (1994), and Schultz (1987, 1989a). For a detailed account of Germany's war aims, see Fischer (1977).

2 CARTOGRAPHY AND NATIONAL TERRITORY AT THE END OF THE FIRST WORLD WAR

42 See, for example, the Guildhall speech of 9 November 1914 by the British Prime Minister Herbert Henry Asquith, in which he mentioned the intent to place the rights of the smaller nationalities "upon an unassailable foundation" (cited after Komarnicki 1957: 12).

43 In the secret Treaty of London (26 April 1915) Italy was promised territories that were clearly not inhabited by Italians, and in the Sykes–Picot Agreement of 1916, France and Britain agreed about the partition of Asiatic Turkey. A secret Franco-Russian Agreement in 1917 even went so far as to promise Russia a free hand in the determination of her western boundaries.

44 On the official recognition of national self-determination, see also Temperley (1969 II: 227, 262, 266).

45 The following types of maps were submitted to Wilson: 1) maps showing boundary claims by governments, subject peoples, and important political groups; 2) maps showing racial and religious composition of the areas based on official statistics; where disputes exist, based on claims; 3) maps showing international trade routes; 4) maps showing spheres of influence, special concessions, protectorates, disguised and avowed (after Gelfand 1976: 344).

46 Thomas Masaryk was a Czech nationalist who later became the president of Czechoslovakia.

47 Incidentally, the eastern limit of the ethnic Polish territory in this map does not extend as far east as the boundary of the Polish state established at the Paris peace conference.

48 The map was translated and reprinted in Oberhummer (1920).

49 Bohemia was the name of choice for the state desired by the Czech nationalists (Masaryk 1917: 2).

50 Masaryk (1917). The text was a reprint of an article in the 22 February 1917 issue of *New Europe*, the map had been supplemented to the British *New Europe* on 25 January 1917.

51 The Teschen district was given to Czechoslovakia at Paris, but transferred to Poland in 1938. Conflicting claims of Poles and Czechs to the Teschen district around the time of the peace conference were later described and illustrated by Witt (1935: 178–9).

52 The territorial claims outlined in this atlas will be discussed in the context of the German critique of Romer's maps in Chapter 3.

53 The maps in memorandum No. 2 and memorandum No. 3 were reprinted in the June 1937 issue of the periodical *Volk und Reich*. Apparently, the Czecho-slovak delegation also submitted a map by the Czech legal historian Jan Kapras, which depicted Germanization activities in Bohemia (Meissner-Hohenmeiss 1937: 436–7).

54 For example, pro-Serb: Ribaric, de Sisic, and Zic (1919); and pro-Italian: *Fiume. Droits et devoirs de l'Italie*, no author, no date. Both pamphlets are preserved at the Memorial Library of the University of Wisconsin-Madison.

55 A. Ischirkoff, a professor of geography, was in charge of the ethnographic maps and V. Zlatarski, a professor of history, assembled the historical maps (Rizoff 1917: XIX).

56 The article appeared in *Patrie serbe*, November–December 1918.

57 The influence of geographers and ethnographic maps on the boundaries in the Balkans is vividly described by Wilkinson (1951). There is also a brief discussion in Taylor (1989: 185–8).

58 For a discussion of geographical arguments for *Mitteleuropa*, see Kost (1988: 269–74), Schultz (1987, 1989a and 1989b).

59 Wilhelm Volz to Foreign Office, 2 July 1923 (AA DiA-2, Nr. 11, Band. 1). See also Volz (W. 1942: 723–4).

60 Penck and Heyde (1919).

61 Schäfer (1919). Note: Häberle (1919: 124) cited this map with the date of 1918.

62 Stahlberg failed to give a complete reference to either map and I have not been able to locate them.

63 The map was entitled *"Répartition de la population polonaise sur le territoire de l'ancienne République de Pologne."* It showed percentages of Polish speakers by district at a scale of 1:2,750,000 and was appended to Freilich (1918).

64 Spett (1918). It is preserved at the Österreichische Nationalbibliothek, Vienna (K III 100.626). It is also reprinted in Geisler (1933).

65 Mehmel (1995: 500) also points out this bias in Penck's map, but does not address its place in the larger context of cartographic revisionism discussed in Herb (1989 and 1993).

66 The Austrian geographer Robert Sieger (1919: 245) pointed out that claims to ethnic German territory in Austria-Hungary came late and therefore faced difficulties in being accepted in the *"übersättigt"* (surfeit) market of neutral publications, such as Swiss books and newspapers. He also lamented the lack of personal contacts to the neutral countries. Therefore, he argued, foreign propaganda was so effective that even purely German areas were wrongly perceived as Slavic abroad. As an example, he mentioned that "the officers of the Entente who came to Klagenfurt . . . were astonished that this Slavic city was so thoroughly German" (Sieger 1919: 245). Sieger was apparently unaware that Klagenfurt was also portrayed as being Slavic in Karl Bernhardi's 1844 map (see n. 17).

67 Much of the cartographic material of the Inquiry Commission is kept at the map collection of the American Geographical Society (AGS) at the University of Wisconsin-Milwaukee. German atlases that are kept there still bear the stamps of the Inquiry or the American Commission to Negotiate Peace.

68 The map, which was presented to the *Reichsamt des Innern*, served primarily as an illustration of a new administrative division of a national German state. In the northeast it was based on the pre-war boundaries of the German Empire, while German Austria in the southeast was delimited along ethnic lines.

69 The text of the treaty is reproduced in its entirety in Temperley (1969 III).

3 A CONCERN FOR ACCURACY

70 Geisler (1926) and Eckert (1925).

71 Geisler (1926, 1933, 1939) also cited maps by Roman Dmowski as examples but his treatment of them was sketchy and consisted mainly of the accusation that Dmowski had deliberately falsified his maps to maximize Polish territorial claims.

72 The problem of the reliability of ethnographic data in the context of the new national boundaries after the Paris peace treaties also stimulated work by statisticians. The Austrian-German Wilhelm Winkler started a *Statistisches Seminar für Bevölkerungs-, Wirtschafts- und Kulturfragen des Grenzlanddeutschtums* at the University of Vienna in the winter of 1921. After he realized that the German ethnic groups in border lands and foreign countries could only be studied in conjunction with other ethnic minorities, he expanded the scope of his research efforts and founded the *Institut für Statistik der Minderheitenvölker an der Universität Wien* in the winter of 1922 (Braunias 1925: 407). The work of the institute consisted of lectures and publications, such as Winkler (1927).

73 As an example, Penck (1921a: 183) referred to the article by Ludwig Bernhard, "Die Fehlerquellen in der Statistik der Nationalitäten," which was included as a preface to Paul Weber, *Die Polen in Oberschlesien*, Berlin 1914.

74 This was admitted in a letter from, the *Reichsminister des Innern* to the *Reichsminister für Wissenschaft, Erziehung und Volksbildung*, 28 March 1940 (BA R153/258).

75 Because of this inconsistency, Geisler's statements will be continuously cross-checked here against the 1916 edition of Romer's atlas.

76 Stahlberg (1921: 10) made a similar statement when he endorsed Dietrich Schäfer's method of distributing bilingual speakers in equal parts between Poles and Germans.

77 "Die Polen," Plate IX. The map shows the distribution of Poles as percentages of the total population. The scale is 1:5,000,000.

78 Paul Langhans, "Die Provinzen Posen und Westpreussen unter besonderer Berücksichtigung der Ansiedlungsgüter und Ansiedlungen . . . nach dem Stande vom 1. Juli 1905," *Deutsche Erde* (1905). Cited after Geisler (1926: fn. 7).

79 The methodological discussions of Polish ethnographic maps also were used as examples in the major textbook of cartography of the time, Max Eckert's two-volume work *Die Kartenwissenschaft* (Eckert 1921 and 1925). Eckert did not examine the key Polish maps themselves, but rather surveyed the treatment they had received by German scientists. In several places, he was extremely critical of Penck's 1921a article. He accused Penck of not having paid enough attention to the dependency between scale and method of representation (Eckert 1925: 463) and pointed to mistakes in Penck's other articles. He gave the example of how Penck misspelled a name and stated that it was wrong of Penck to use the term isarithm to designate Romer's method. However, Romer had used precisely this word in the commentary to plate IX of his 1916 atlas. Eckert's rather harsh and unfounded criticism of Penck raises the question whether there were personal frictions between these two scientists, who were both important figures in German cartography.

80 Penck did not give any further reference to the map.

81 Foreign Office to Generaldirektor Brackmann (*Preußische Staatsarchive*) and to the *Wirtschaftspolitische Gesellschaft*, Berlin, 12 February 1934 (BA R153/1114). See also Brackmann to Regierungspräsident Budding, Marienwerder, 13 December 1933 (BA R153/708).

82 The maps accompanying the monograph included: 1) a reprint of Spett's map; 2) a "correct" version of a language map for the same territory which used the same method of representation, areal coloring, but also designated the languages of the Kaschubes, the Masures, and other linguistic sub-groups (for a very simplified later reproduction of this map, see Schadewaldt 1939a: 235); 3) a map portraying the deviations between the first two maps; 4) a map of the plebiscite results in Upper Silesia; 5) a map of the plebiscite in East and West Prussia.

83 Foreign Office to Generaldirektor Brackmann (*Preußische Staatsarchive*) and to the *Wirtschaftspolitische Gesellschaft*, Berlin, 12 February 1934 (BA R153/1114).

84 Foreign Office to Generaldirektor Brackmann (*Preußische Staatsarchive*) and to the *Wirtschaftspolitische Gesellschaft*, Berlin, 12 February 1934 (BA R153/1114).

85 Generaldirektor Papritz (*Publikationsstelle-Berlin*) to Regierungspräsident Budding, Marienwerder, 13 March 1934 (BA R153/1114).

86 Perthes Verlag to Dr Loessner (*Deutsche Gesellschaft zum Studium Osteuropas*), 14 December 1933 (BA R153/1114). Geisler (1933: preface). The map analysis was published as part of the Perthes monograph series *Petermanns geographische Mitteilungen, Ergänzungshefte*.

87 With the exception of one incident – the commission of a map of the Upper

Silesian plebiscite results in 1921 – the Foreign Office always responded either to suggestions by private groups or to foreign cartographic activities.

88 Foreign Office to Generaldirektor Brackmann (*Preußische Staatsarchive*) and to the *Wirtschaftspolitische Gesellschaft*, Berlin, 12 February 1934 (BA R153/1114).

89 Regierungspräsident Budding to Generaldirektor Brackmann (*Preußische Staatsarchive*), 2 August 1935, and copies of correspondence by Budding to Ministry of the Interior (BA R153/436).

90 The map was prepared by Herbert Heyde based on Penck's design (Penck and Heyde 1921).

91 In contrast to his small-scale color map in the *Zeitschrift der Gesellschaft für Erdkunde zu Berlin* (Penck 1919c), the color choice for his 1:300,000 map was more objective: he chose blue for Germans, red for Poles, purple for Kaschubes, orange for Bilinguals, and green for Speakers of Other Languages (Penck 1921a).

92 Plankammerinspektor Gerke, *Kartographische Darstellung der Abstimmungsergebnisse von Oberschlesien*, 1:100,000. Plankammer des Preußischen Statistischen Landesamtes, 1921. Cited after Stahlberg (1921: 21).

93 Leo Wittschell, "Die völkischen Verhältnisse in Masuren und dem südlichen Ermland," *Veröffentlichungen des Geographischen Instituts der Universität Königsberg*, Hamburg 1926, Heft 5. Cited after Geisler (1933: 52).

94 The title of the 1922 map of Upper Silesia was Volz-Rosenberger, "Besiedlungskarte von Oberschlesien," *Veröffentlichungen der Schlesischen Gesellschaft für Erdkunde*, vol. 3, Breslau 1922. Cited after: Zur Bevölkerungskarte . . . (1925: 160).

95 The map was an initiative of Wilhelm Volz (see Chapter 4).

96 Maps by Geisler using this method also appeared in Erich Keyser, ed. *Der Kampf um die Weichsel. Untersuchungen zur Geschichte des polnischen Korridors*, Stuttgart–Berlin–Leipzig: Deutsche Verlagsanstalt, 1926. Cited after Ziegfeld (1926: 720).

97 It is interesting to note that Geisler's nationalistically motivated idea to base political boundaries on small natural regions has found new currency – albeit in the opposite political corner – in the environmental movement's call for ecologically based administrative units.

98 That is, whether different nationalities were employers or employees.

99 The plebiscite result in East Prussia was so clear – more than 90 per cent voted pro-German – that it did not spark a cartographic feud.

100 The maps are cited after Stahlberg (1921: 20–1), who did not provide any further references.

101 Plankammerinspektor Gerke, *Kartographische Darstellung der Abstimmungsergebnisse von Oberschlesien*, 1:100,000. Plankammer des Preußischen Statistischen Landesamtes, 1921. Cited after Stahlberg (1921: 21).

102 Penck and Heyde (1919).

103 Foreign Office, internal memorandum, "Gesichtspunkte zur Behandlung der Oberschlesischen Frage nach der Abstimmung," dated 19 March 1921 (AA Deutschland 8, Bd. 2).

4 NEW CONCEPTS OF NATIONAL TERRITORY

104 K. C. von Loesch to Foreign Office, 12 November 1921 (AA DiA-2 Nr. 11, Bd. 1).

105 A. Murphy's (1990: 536) statement that the territorial claims at Paris "centered around historical considerations," appears questionable in the light of the present research (see Chapter 2).

106 His main goal was to combat the legalistic concept of the state, which regarded the state as being defined through its constitution (Kristof 1960).

107 F. Metz to A. Brackmann, 27 September 1939 (BA R153/290).

108 Loesch to Foreign Office, 13 November 1921 (AA DiA-2, Nr. 11, Bd. 1).

109 "Karte 2: Die siedlungsgeographischen Grundlagen der Verteilung von Deutschen und Polen." The other two maps were: "Karte 1: Verteilung der deutschen und polnischen Stimmen bei der Abstimmung;" and "Karte 3: Die Verbreitung deutsch-stimmender Oberschlesier polnischer Zunge."

110 The final boundaries became effective 20 October 1921. Ten copies of Volz's work arrived at the *Reichskanzlei* on 22 November 1921, and ten copies of Dietrich's on 24 October 1921 (BA R43I/360 and 361).

111 Volz never published a map of his conception of the German *Volksboden*.

112 There is no agreement in the literature on this issue. Mosse (1981) and Fahlbusch (1994) consider Ratzel to be a proponent of *völkisch* ideas.

113 Boehm was a leading neo-conservative thinker who provided links between several neo-conservative organizations, such as the *Juni-Klub*, the *Deutscher Schutzbund*, and the *Deutsche Rundschau*. Neo-conservatives, the proponents of the radical non-Nazi German Right, can be considered the intellectual trailblazers of the *völkisch* movement. For a brief summary of neo-conservatism, see Stark (1981: 4–9).

114 This is expressed most forcefully in Kjellén (1914), which went through nineteen editions in Germany in 1914–18. After the First World War it was issued under the title *Die Grossmächte und die Weltkrise* in two more editions. See also Kristof (1960) and Sieger (1924). For a comprehensive analysis of Kjellén, see Holdar (1992).

115 I would like to thank Marjorie Lamberti (Middlebury College) for bringing my attention to this issue.

116 On the development of suggestive cartography and Ziegfeld, see Chapter 6.

117 For a discussion of Penck's wall map, see Chapter 7.

118 On the issue of German defeat and sense of self-worth, see also Jaworski (1978: 382).

5 COORDINATION OF CARTOGRAPHIC REVISIONISM

119 *Ergebnis der kommissarischen Beratung über die wissenschaftliche Bearbeitung der Grenzzerreissungsschäden im Osten vom 18. 6. 1929 im Reichsministerium des Innern* (BA R43I/1800, folio 21).

120 There were a variety of specialized research institutes involved in work on German national territory, such as the *Deutsches Ausland-Institut* in Stuttgart and the *Institut für Grenz- und Auslandstudien* in Berlin, but the only coordinating center was the *Stiftung*.

121 Before Fahlbusch's (1994) meticulously researched work, little was known about the *Stiftung*. Burleigh (1988: 25) dated it to 1926 and attributed the initiative to the German Interior Ministry; Roessler (1990: 53–4) mentioned 1923 as the foundation date and Penck, Loesch, and the Interior Ministry as key figures. D. Murphy (1992: 285–6) dated it to 1922 and cited a source which claimed it was a creation of the Interior Ministry. Fahlbusch (1994: 63) also mentions a "*Grenzmarkenausschuß*" as a precursor of the *Stiftung*.

122 W. Volz to Foreign Office, 15 November 1921; K. C. von Loesch to Foreign Office, 18 November 1921 (AA DiA-2, Nr. 11, Bd. 1).

123 The *Deutscher Schutzbund* was one of the key centers of the neo-conservative movement (Mauersberger 1971: 30). On the *Deutscher Schutzbund* and Loesch, who was also a member of the *Juni-Klub*, see Mauersberger (1971: 41–6).

124 Fahlbusch (1994: 58–9) argues that the motivation for this cooperation between *völkisch* nationalists and geographers was a resolution by the association of German geographers which demanded that Germany's pre-war boundaries always be included in maps.

125 K. C. von Loesch to Foreign Office, 18 November 1921 (AA DiA-2, Nr. 11, Bd. 1).

126 W. Volz to Foreign Office, 15 November 1921 (AA DiA-2, Nr. 11, Bd. 1).

127 The list was compiled from the following documents: W. Volz to Foreign Office, 15 November 1921 and 12 December 1921 (AA DiA-2, Nr. 11, Bd. 1).

128 W. Volz to Foreign Office, 15 November 1921 and 12 December 1921. *Denkschrift über die Notwendigkeit einer wissenschaftlichen Erfassung des Ostdeutschtums; Bericht über die am Dienstag, den 23. August 1921 im Landeshaus zu Breslau stattgefundene vertrauliche, informatorische Besprechung über die Notwendigkeit und Durchführung einer wissenschaftlichen Erforschung des Grenz- und Auslandsdeutschtums* (AA DiA-2, Nr. 11, Bd. 1).

129 Handwritten notes on letter from K. von Loesch to Foreign Office, 18 November 1921 (AA DiA-2, Nr. 11, Bd. 1).

130 *Bericht über die am Dienstag, den 23. August 1921 im Landeshaus zu Breslau stattgefundene vertrauliche, informatorische Besprechung über die Notwendigkeit und Durchführung einer wissenschaftlichen Erforschung des Grenz- und Auslandsdeutschtums*, p. 4 (AA DiA-2, Nr. 11, Bd. 1).

131 K. von Loesch, *Bericht über die wissenschaftlich-politische Tagung*. The report bears a stamp with the date "17 January 1922" (AA DiA-2, Nr. 11, Bd. 1). The following discussion on the January 1922 meeting is based on the same document.

132 Partsch was a professor of law at the University of Bonn. It is not clear if he was related to the famous geographer Joseph Partsch, who was a professor at the University of Breslau.

133 K. von Loesch, *Bericht über die wissenschaftlich-politische Tagung*. The report bears a stamp with the date "17 January 1922" (AA DiA-2, Nr. 11, Bd. 1).

134 K. von Loesch, *Bericht über die wissenschaftlich-politische Tagung*, Anlage 4. The report bears a stamp with the date "17 January 1922" (AA DiA-2, Nr. 11, Bd. 1).

135 The following discussion of the meetings is compiled from the conference reports (*Tagungsberichte*) in AA DiA-2, Nr. 11, Bd. 1–8.

136 "*Übersichtskarte des wendischen Sprachgebiets und Spezial-Verkehrskarte der Ober- und Niederlausitz*. Reproduktion der im Jahre 1886 herausgegebenen wendischen Karte von Oberstudienrat Dr. Muke, Schleife 1919." Cited after: Zur Bevölkerungskarte . . . (1925: 158).

137 K. von Loesch to W. Volz, 1 October 1924 (AA DiA-2, Nr. 11, Bd. 2).

138 Interior Ministry to Foreign Office, 16 October 1924 (AA DiA-2, Nr. 11, Bd. 2).

139 *Bevölkerungskarte der Ober- und Niederlausitz auf Grund der Volkszählung von 1910*, 1:200,000, no author (Zur Bevölkerungskarte . . . 1925). Volz also initiated a reduced version of the map at a scale of 1:400,000, which used Penck's absolute representation method (W. Volz to *Statistisches Reichsamt*, 26 July 1926, AA DiA-2, Nr. 11, Bd. 3).

140 Penck's statement at the Marienburg meeting was a reaffirmation of a resolution passed a few months earlier by the *Deutscher Geographentag* (see Chapter 7).

141 Interior Ministry to Foreign Office, 25 June 1926 (AA DiA-2, Nr. 11, Bd. 3).

142 *Niederschrift über die kommissarische Besprechung vom 15. Juli 1927* (AA DiA-2, Nr. 11, Bd. 4).

143 The article was not identified, but it was probably Jäger (1924).

144 Report on the meeting of the *Beirat* of the *Stiftung* on 1 December 1926, page VI (AA DiA-2, Nr. 11, Bd. 4).

145 The conference (*"kommissarische Besprechung"*) of 15 July 1927 included members of the following agencies: *Reichsministerium des Innern, Auswärtiges Amt, Reichsverkehrsministerium, Statistisches Reichsamt, Preußisches Ministerium des Innern, Preußisches Kultusministerium, Preußisches Ministerium für Volkswohlfahrt, Preußisches Landwirtschaftsministerium.*

146 *Ergebnis der kommissarischen Beratung über die wissenschaftliche Bearbeitung der Grenzzerreissungsschäden im Osten vom 18. 6. 1929 im Reichsministerium des Innern* (BA R43I/1800, folio 21).

147 Schwalm was the "chief scientific secretary" (*Erster Wissenschaftlicher Sekretär*) of the *Stiftung.*

148 Report on the meeting of the *Verwaltungsrat* of the *Stiftung* on 10 October 1929 (AA DiA-2, Volz-Stiftung, Bd. 2.).

149 Karte 1: *Einzugsbereiche der offenen Ladengeschäfte und des Handwerks 1913.* Karte 1a: *Einzugsbereiche der offenen Ladengeschäfte und des Handwerks (Teilstück der Ostpr. Grenze) 1913 und 1929.*

150 A variety of editors were commissioned by the *Stiftung* for the *Handwörterbuch* project, among them Emil Meynen.

151 The only references given were: *"Heimatatlas von Sachsen"* and *"Bayerischer Sprachatlas." Etat der Stiftung für das Rechnungsjahr 1930* (AA DiA-2, Volz-Stiftung, Bd. 3).

152 Albin Oberschall, *Die Nationalitätenfrage in der Tschechoslowakei (Mit Sprachenkarte)*, Reichenberg: Sudetendeutscher Verlag Hans Kraus, 1927 (AA DiA-2, Nr. 11, Bd. 7).

153 Jakob Bleyer, *Deutsche Siedlungen in Rumpf-Ungarn, 1920*, 1:400,000, Budapest: Kartographische Kunstanstalt Georg Klösz & Sohn, 1928. The *Stiftung* even carried out the final draft of Bleyer's map and had it sent to him secretly through diplomatic channels (AA DiA-2, Nr. 11, Bd. 8). Copies of Bleyer's map are kept at the Bundesarchiv (BA DAI-Kart, No. 204 and No. 205).

154 The plan for the Spek map was mentioned in a report appended to a letter by the *Stiftung* to the Foreign Office, 8 October 1930 (AA DiA-2, Volz-Stiftung, Bd. 4). In 1923, Spek had published a map through the *Kulturamt des Verbandes der Deutschen in Groß-Rumänien* entitled *Karte der deutschen Siedlungen Groß-Rumäniens,* 1:500,000 (BA DAI-Kart, No. 163).

155 For a discussion of this atlas project, see Oberkrome (1993: 84–7), who notes that the work was also intended to mark the limits of the German *Volksboden.* However, it seems that *Volksboden* in this context was meant to encompass only the German language area.

156 AA DiA-2, Nr. 11, Bd. 6–7.

157 The institute was to be headed by Heinrich Schmidt. Its focus was to be the *Donauschwaben (Stiftung* to Foreign Office, 2 October 1930. AA DiA-2, Volz-Stiftung, Bd. 3).

158 Memorandum by W. Volz, dated 23 July 1926, p. 10 (AA DiA-2, Nr. 11, Bd. 3). It is not clear whether Volz was referring to the Geneva negotiations preceding the protocols of 2 October 1924 (peaceful settlement of international disputes) or of 17 July 1925 (international agreement on the ban of chemical and biological weapons).

159 Notes by von Heeren on the meeting of the *Mittelstelle* on 17 January 1925

(AA DiA-2, Nr. 11, Bd. 2) and copy of correspondence from *Statistisches Reichsamt* to *Mittelstelle*, 8 October 1926 (AA DiA-2, Nr. 11, Bd. 4).

160 *Ergebnis der kommissarischen Beratung über die wissenschaftliche Bearbeitung der Grenzzerreissungsschäden im Osten vom 18. 6. 1929 im Reichsministerium des Innern* (BA R43I/1800, folio 21).

161 Memorandum by Volz, dated 27 March 1925, and appended statement by Loesch (AA DiA-2, Nr. 11, Bd. 2).

162 AA DiA-2, Nr. 11, Bd. 3.

163 The three maps were: 1) *Bevölkerungskarte von Oberschlesien (ehemaliges Abstimmungsgebiet) auf Grund der Reichstagswahl vom 7. 12. 1924*, 1:400,000; 2) *Bevölkerungskarte der Ober- und Niederlausitz auf Grund der Volkszählung von 1910*, 1:200,000; 3) *Bevölkerungskarte der Ober- und Niederlausitz auf Grund der Volkszählung von 1910 (Punktkarte)*, 1:400,000.

164 W. Volz to *Statistisches Reichsamt*, 26 July 1926 (AA DiA-2, Nr. 11, Bd. 3).

165 AA DiA-2, Nr. 11, Bd. 3. The 1925 census results for the border regions of Germany – albeit not extending beyond the contemporary political boundaries – were published in an atlas (*Reichszentrale für Heimatdienst* 1929). The maps, which used Penck's absolute representation method, were prepared by the *Preußisches Statistisches Landesamt*, the same institution which had prepared the map of the Upper Silesian plebiscite results for the Foreign Office in 1921. It seems that the intention of the atlas was to contradict further Polish claims to existing German territories. In a June 1929 meeting, a government official expressed concern that Poland was vigorously working on substantiating claims to more German territory (*Ergebnis der kommissarischen Beratung über die wissenschaftliche Bearbeitung der Grenzzerreissungsschäden im Osten vom 18. 6. 1929 im Reichsministerium des Innern*; BA R43I/1800, folio 21).

166 Untitled memorandum by F. Metz, ca. 1928 (AA DiA-2, Nr. 11, Bd. 8).

167 Circular letter by A. Penck to the editors of the *Handwörterbuch*, 20 November 1928; W. Volz to A. Penck, 1 December 1928 (AA DiA-2, Nr. 11, Bd. 8).

168 There is even more uncertainty in the literature about the end of the *Stiftung* than about its origins. Fahlbusch (1994: 96–7) states that the *Stiftung* was only formally dissolved on 18 July 1940. However, I found no records of activity by the *Stiftung* after August 1931. The information about the dissolution on 8 August 1931 is taken from a newspaper clipping from *Der Montag Morgen* (AA DiA-2, Volz-Stiftung, Bd. 5). See also Voigt (1965: 381). There was also a central office for research on Eastern questions (*Zentralstelle für Ostforschung*) which had been founded at the suggestion of Volz in 1928, and which was to be closely associated with the *Stiftung*. It was dissolved together with the *Stiftung* (Burleigh 1988: 70–1). Roessler (1990: 55) cites a letter by Brackmann to Penck which dates the dissolution of the *Stiftung* as 19 January 1932.

169 H. Hassinger to Interior Ministry, 10 November 1931; R. Gradmann to Interior Ministry, 23 December 1931; Interior Ministry to Metz, 29 December 1931 (AA DiA-2, Volz-Stiftung, Bd. 5). See also Fahlbusch (1994: 95–7).

170 The *Volkswissenschaftlicher Arbeitskreis* was an organization of the *Verein für das Deutschtum im Ausland* and was affiliated with the *Deutsches Ausland-Institut*. For a discussion of the *Forschungsgemeinschaften*, see Roessler (1990, ch. 4.3.3) and Burleigh (1988).

171 After the dissolution of the *Stiftung*, Volz faded from the political scene. During the Nazi period, he faced difficulties because his wife was Jewish (Sandner 1983: 76–8); this was also stated by Richard Hartshorne in an interview with the author.

6 MAPS AS WEAPONS

172 See for example Troll's (1947) attempt to whitewash the involvement of geographers. Recent studies have revealed the depth of the complicity of geographers with National Socialism. For a survey of the German literature on the subject, see Sandner (1988).

173 For a comprehensive treatment of *Geopolitik* and its relationship to political geography, see Kost (1988). The usefulness of such geo-determinist arguments was already recognized during the war in the formulation of war aims by German geographers (see Schultz 1987).

174 On Haushofer, see Jacobsen (1979) and Diner (1984).

175 Karl Haushofer in a letter to his wife Martha, 4 June 1917. Cited by Jacobsen (1979 I: 484).

176 Earlier names of the VDA were: *Deutscher Schulverein* (from its foundation in 1881 until 1887) and *Allgemeiner Deutscher Schulverein zur Erhaltung des Deutschtums im Ausland* (1887–1908). For a general description of the VDA, see Possekel (1986). The *Alldeutscher Verband* was founded in 1891 (Jaworski 1978: 371).

177 *Völkisch* nationalists, such as Karl C. von Loesch and Albert Wacker, were constituent members of the *Stiftung für deutsche Volks- und Kulturbodenforschung*, which coordinated and supported the development of such maps (see Chapter 5).

178 E.g. Josef März, *Landmächte und Seemächte*, Berlin, 1928 (Weltpolitische Bücherei X).

179 The original maps are on pp. 128–9 and 144–5 in Stuart (1920).

180 According to Fahlbusch (1994: 59, fn. 123) it was the VDA which had voiced the demand before.

181 Karl Haushofer, Die suggestive Karte. *Grenzboten* 1 (1922): 17–19. The article was reprinted in Haushofer (1928).

182 Pickles (1992: 203) states that Goebbels thought along similar lines and always included a nucleus of truth in his propaganda campaigns. As Welch has pointed out repeatedly (1983: 2; 1993: 5), it is a common misconception that "propaganda consists only of lies and falsehood."

183 There was one article on suggestive cartography in the May/June 1927 issue of *Volk und Reich* (Ziegfeld 1935: 244), but this was an exception. The periodical was geared toward the practical application of suggestive maps, not discussion.

184 AA DiA-2, Nr. 11, Bd. 2. BDC, personal file A. H. Ziegfeld.

185 BDC, personal file A. H. Ziegfeld. Black-white-red were the colors of the imperial German flag, which was used by many conservatives to show that they rejected the Weimar system and its official colors (black-red-gold).

186 For example, Ziegfeld published a book together with Hans Christoph Ade, *Pioniere im Osten*, Stuttgart: Union Deutsche Verlagsgesellschaft, 1923. In addition, Ziegfeld was familiar with England. He went to England to study at Oxford University in 1913, but was interned for the duration of the First World War because he was a foreigner. Albrecht Penck, who was surprised by the outbreak of the war when he returned to Germany from a research trip, also was interned in England during the First World War. A striking coincidence since both of them were instrumental in the politicization of the cartography of German national territory: Ziegfeld initiated the development of suggestive representations and Penck was involved in questions of accuracy and the coordination of research (see Chapters 4 and 5).

187 BDC, personal file A. H. Ziegfeld.

188 Heft 2, Blatt 9, "Speicher und Nahrung." Heft 3, Blatt 15, "Wasser und Weg." Heft 3, Blatt 16, "Niederer und erhabener Sitz." Like all the other authors of articles, Ziegfeld provided the information for the drawing of the maps, but L. Ritter von Wilm was the cartographer who executed the numerous maps of the atlas in a consistent design.

189 The sub-title of the map read: "Five and a half million Germans are to be separated from the Empire in the east and west!" Copies of the poster-size map are preserved at the BA (Poster 2/7/73) and the AA (R 23076).

190 A copy of the publication is kept at the BfPB. Most publications of the *Reichszentrale* had a print run of at least 30,000 copies, which were distributed to its 30,000 spokespersons who gave lectures for the *Reichszentrale*.

191 AA Presseabteilung, RfH, Deutschland 19, Geheim, Bd. 3.

192 For an example of pincer-like arrows converging on the province, see Wittschell (1926: 430).

193 Wippermann (1976: 105, fn. 28). See also letter from *Reichszentrale* to *Presseabteilung der Reichsregierung*, 19 June 1923, p. 2 (AA Presseabteilung, RfH, Deutschland 19, Geheim, Bd. 6).

194 The memorandum by Otto Brandt, entitled *Amtliche deutsche Werbearbeit im Auslande*, Berlin: W. Büxenstein, 1924, is included in AA Abt. Inland, Ref. Deutschland, Po. 26, Politische und kulturelle Propaganda, Bd. 1.

195 Volz's article is obviously biased. He is trying to present his own accomplishments in the best possible way. But in the light of the aforementioned documentation of the lack of attention to maps in official propaganda, Volz's statement is still useful as corroborative evidence.

196 For example, Edwin Runge Verlag, Vowinckel Verlag.

197 Wippermann (1976: 171–235). See also the secret report attached to a letter from the *Reichszentrale* to *Presseabteilung der Reichsregierung*, 9 February 1923 (AA, Presseabteilung, RfH, Deutschland 19, Bd. 1). For a partial list of the associated *völkisch* associations, see Mauersberger (1971: 45–6).

198 The *Wirtschaftspolitische Gesellschaft* was founded in 1922 and supported by industries of the Rhine region and Westphalia. It's goal was the education of foreigners about the Treaty of Versailles. It disseminated information material and organized and supported excursions. One participant of such an excursion was the American geographer Richard Hartshorne, who traveled to Upper Silesia (*Wirtschaftspolitische Gesellschaft, Tätigkeitbericht Januar 1931–15. März 1933*; BA R153/80).

199 *Volk und Reich* was almost exclusively supported by German industry (*Wirtschaft*) (F. Heiss to Foreign Office, 21 July 1933; AA DiA-2, Volk und Reich, Bd. 1).

200 BA R57neu/1005, Nr. 2; AA Presseabteilung, Deutschland 6, Volk und Reich, Band 1. In fact, *Volk und Reich* and the *Deutscher Schutzbund* had the same address: Motzstrasse 22, Berlin. The *Deutsche Rundschau*, which was edited by the neo-conservative Rudolf Pechel, and the *Juni-Klub* were also housed there (Mauersberger 1971: 37, 41). According to a report by German scientists submitted to British Intelligence after the Second World War, the journal *Volk und Reich* was founded in 1925 by K. von Loesch (head of the *Deutscher Schutzbund*) and Martin Spahn (professor at the *Hochschule für Politik*, Berlin); PRO 1031/140, p. 147.

201 BA R153/80.

202 BDC, personal file A. H. Ziegfeld.

203 *Zehn Jahre Deutscher Schutzbund, 1919–1929*, p. 29. This publication is included in AA DiA-2, Deutscher Schutzbund, Bd. 9.

204 Copy of a letter of recommendation for Ziegfeld by von Wrangel and von Loesch of the *Deutscher Schutzbund*, dated 30 June 1928 (BDC, personal file A. H. Ziegfeld).

205 Ziegfeld's father, for example, organized regular German musical events in Japan (BDC, personal file A. H. Ziegfeld).

206 In an undated fragment of a personal report in his BDC file, Ziegfeld indicates that he left the *Schutzbund* because it had become increasingly dependent on the "former regime." He went on to claim that he left the "contaminated air" (*unlautere Luft*) of the *Schutzbund* to work with other people engaged in "national political" (*volkspolitische*) activities such as the journal *Volk und Reich*. He also mentioned in a letter to the Foreign Office (received on 8 December 1933) that the older generation in the executive committee of the *Deutscher Schutzbund* resisted the activities of the younger generation, such as himself.

207 Internal memo dated 11 July 1929; Press Office of the Foreign Office to von Grünau, 16 July 1929 (AA, Presseabteilung, Deutschland 6, Volk und Reich, Bd. 1).

208 Copy of a recommendation (*Gutachten*) for Ziegfeld by Haushofer, dated 15 November 1925 (BDC, personal file A. H. Ziegfeld).

209 The *Deutscher Klub* also brought him in contact with another important suggestive cartographer who worked for the VDA, Friedrich Lange. Lange, Boehm, von Loesch, and Ziegfeld were some of the founding members of the *Deutscher Klub*. On Hermann Ullmann, see also Chapter 7.

210 Albrecht Haushofer, the son of Karl Haushofer, was even directly involved in the journal *Volk und Reich*. He edited a regular column starting in 1926 (Korinman 1990: 276).

211 Dissemination of works containing suggestive maps was also encouraged through the *Grenzbüchereidienst*, which supplied *völkisch* organizations, scientific institutes, libraries, and other institutions, particularly in remote border regions, with recent publications on German issues, such as the journal *Volk und Reich* (BA R57neu/1005, Nr. 19).

212 Braun and Ziegfeld (1927) was the first suggestive atlas, which is a further indication of Ziegfeld's innovative and influential role in the propagation of suggestive cartography. No earlier atlas was mentioned in any of the surveys on suggestive maps, such as Ziegfeld (1925 and 1926) and Schumacher (1935). Haushofer (1932: 735) mentioned Schmidt and Haack (1929) as the first suggestive atlas, but failed to notice that the first edition of the Braun and Ziegfeld atlas had appeared in 1927, that is, two years earlier than the first edition of Schmidt and Haack.

213 *VDA Jahresbericht 1928*, p. 101 (BA ZSg.1–142/19).

214 AA DiA-2, VDA, Bd. 6. A comparative map postcard by the *Arbeitsgemeinschaft* which depicts corridors through Germany and Great Britain is preserved at the Bundesarchiv (BA R153/80).

215 For example, Braun and Ziegfeld (1927); Schmidt and Haack (1929); Knieper (1931); Springenschmid (1934); and Czajka et al. (1933).

216 This was stressed in the preface to Schmidt and Haack (1929).

217 E.g. Vowinckel Verlag, Edwin Runge Verlag, and the Zentralverlag in co-operation with the Deutscher Lichtbilddienst.

218 *Gesamtverzeichnis 1939, Volk und Reich*, p. 43 (BA R153/1190).

219 The price for an annual subscription was RM 20, individual issues were available for RM 3. (From an advertisement in *Der Heimatdienst*, November 1929: 395.)

220 For example, he trained Dora Nadge who worked under him at *Volk und Reich*

and advised Kurt Trampler on the cartography for Trampler (1935). Ziegfeld's sign was a capital H in which a small circle topped by a cross was inserted in the upper opening.

221 Nadge, the only female suggestive cartographer, was a graphic designer by training and not a psychologist as Schumacher (1935: 250) believed (BDC, personal file Dora Nadge).

222 The handwriting style had been developed by the graphic designer Ludwig Sütterlin around the turn of the century. For an example of Springenschmid's maps, see fig. 6.4.

223 Springenschmid was a teacher by training who lost his position in Austria because of his illegal membership in the Austrian Nazi party. He became a member of the NSDAP Austria in 1931 and of the NSLB Austria on 1 October 1932. He subsequently became active in National Socialist indoctrination (*Schulung*) and advanced to *Gauschulungsleiter* and director of the *Abteilung II (Erziehung und Volksbildung)* of the office of the *Reichsstatthalter* Rainer in Salzburg. Springenschmid also had a career as a member of the SS, where he achieved the rank of SS-*Hauptsturmbannführer* (BDC, personal file K. Springenschmid).

224 The full title was *Sprachenkarte von Mitteleuropa (Von Triest bis Trollhättan, von Dünkirchen bis Dünaburg und Konstanza)*; it appeared in two editions (78 × 98 cm and 39 × 49 cm). Other maps by Lange were: *Das neue Europa* (1:4,000,000) and *Das neue Deutschland* (1:1,000,000). All were published in Berlin by Dietrich Reimer.

7 CREATING A UNIFIED MESSAGE

225 Statement by Wilhelm Volz quoted in the report on the Witzenhausen/Kassel conference, March 1924 (AA DiA-2, Nr. 11, Bd. 2).

226 See also the references in Geisler (1933: 12, fn. 1).

227 The DKG was a private association founded in 1937, which acted as a liaison office between the government and private cartographic organizations (see also Chapter 9).

228 Dehmel, 5,2: Tafel 1 (WWA).

229 A complete collection of Diercke atlas editions is preserved at the WWA.

230 The process of cartographic Germanization continued after 1933. See Chapter 9, n. 432.

231 Up to the 42nd edition (1920), the map was entitled *Völkerkarte von Mittel- und Südosteuropa*, the title in the 47th to 50th editions in 1926–31 was: *Völker- und Sprachenkarte des mittleren Europa am Ende des 19. Jahrhunderts*. The scale was 1:15,000,000 throughout. An incomplete collection of the Putzger atlas editions is preserved at the GEI. Wolf (1991: 27–8) states that this map, which was first introduced in 1900, was not used for the Austrian edition, apparently with the intention not to unsettle schoolchildren in the Austro-Hungarian monarchy with the conflicts among national groups.

232 The following discussion of the Westermann wall maps is based on the correspondence between the map author and the Westermann Verlag (WWA, v30/217). Since privacy rights of the map author are still in effect, the real name of the map author is substituted by the fictitious initial K.

233 The building officer had a personal interest in the lost eastern territories. His letterhead revealed that he had owned a construction business in Bromberg, Poznan. Most of the district of Poznan, including Bromberg, was given to Poland in the Treaty of Versailles.

234 The *Deutscher Schutzbund* was founded in 1919 to coordinate plebiscite preparations. It was also the parent organization of all groups representing German interests abroad (see also Chapter 6).

235 An official from this institution, *Plankammerinspektor* Gerke, had prepared a map of the Upper Silesian plebiscite results in 1921 (see Chapter 3).

236 The demands of the VDA were circulated in a letter to all German map and atlas publishers. Among geographers it only seems to have been publicly endorsed by Paul Langhans, who was a professed *völkisch* nationalist (Langhans 1916).

237 The map on page 18a of the atlas was at a scale of 1:5,000,000 (Ziegfeld 1925: 444). On other maps in the atlas which depicted parts of Czechoslovakia as well as on a separate ethnographic map of Czechoslovakia, German place names were given preference (Jaworski 1980: 252).

238 On the role of the DNVP in Weimar, see Mosse (1981: 238–53).

239 There is some confusion in the article regarding the edition of *Meyers Konversationslexikon*. First it mentioned the 7th edition and later the 5th edition. However, since Ziegfeld (1926: 718) praised the 7th edition as being up to date on current research, it appears most likely that the *Ostdeutsche Morgenpost* meant the 5th edition. The title of the article in the *Ostdeutsche Morgenpost* is "Die Wahrheit über Oberschlesien." A clipping of the article is included in AA DiA-2, Nr. 11, Bd. 4.

240 AA DiA-2, Nr. 11, Bd. 4.

241 Report on the meeting of the *Beirat* of the *Stiftung*, 1 December 1926, p. 4 (AA DiA-2, Nr. 11, Bd. 4).

242 *Stiftung* to Velhagen & Klasing, 1 February 1927 (AA DiA-2, Nr. 11, Bd. 4).

243 *Bericht über eine Besprechung mit dem Mitarbeiterkreis des Seydlitz'schen Lehrbuches ...* (AA DiA-2, Volz-Stiftung, Bd. 3).

244 Uhlig (1919: 284). Uhlig listed more than fifteen such maps.

245 Steinacher (VDA) to Foreign Office, 14 December 1936 (AA DiA-2, VDA, Bd. 16). Objections to the map were also raised in the United States of America – the map was reprinted in the 17 March 1935 edition of the *New York Times*. Humbel (1977: 152–3) mentions that criticism about this map was raised in Switzerland as late as 1937.

246 Memo dated 13 June 1929 (AA Abt. II (Politik 2), Dt.-österreichischer Volksbund – Österreich, Bd. 1).

247 This was stated on the map. There are two copies of the map at the Bundesarchiv: BA-Kart 861/5; BA-DAI Kart Nr. 240.

248 "Der deutsche Volks- und Kulturboden," *Deutsche Arbeit*, vol. 25, November 1925: 87–8. The price for the atlas insert map in 1928 was RM 0.15 (*Das Auslandsdeutschtum im Lehrplan*, 7. Ausgabe, April 1928, p. 21). According to Meynen (1935: 19, fn. 6), the atlas map was at a scale of 1:16,000,000.

249 Scheer (1927); *Putzgers Historischer Schulatlas* (1931); *Diercke Schulatlas* (1935); *Columbus Weltatlas* (1937).

250 For a reproduction of another modified version of the Penck/Fischer map in Otto Maull's 1933 regional geography of Germany, see Schultz (1990: 66).

251 Cited after Wolf (1978: 709).

252 According to a newspaper clip from the *Berliner Zeitung*, 7 January 1933, which is included in BA R153/71, the center point of the exhibition was a series of maps. Among them were a map which showed the disruption of railroad connections and a diagram which revealed the importance of the corridor as a market area for East Prussia before the war.

253 For a representative example of such a map comparison, see Ziegfeld and Kries (1933: 351).

254 For example, the *Landeshauptleute* of the provinces Ostpreußen, Grenzmark Posen-Westpreußen, Pommern, Brandenburg, Niederschlesien, and Ober-schlesien sent a memorandum with maps, entitled *Die Not der preußischen Ostprovinzen* (Königsberg 1930) to an office of the federal government (*Reichskanzlei*) and asked for assistance (BA R431/1800).

255 The map in Werner which depicted the dissection of the Upper Silesian industrial region was also included in the memorandum *Die Not der preußischen Ostprovinzen*, mentioned above.

256 The maps in question were included in *Seidlitz Geographie*, Heft 4, and in Walter Gehl, *Geschichte für höhere Schulen*, Heft 4.

257 Volz to Census Office, 29 July 1926 (AA, DiA-2, Nr. 11, Bd. 3).

258 Herbert Heyde, *Bericht über den 24. Deutschen Geographentag in Danzig, Pfingsten 1931*, p. 7 (WWA 30/3). Heyde had prepared maps in collaboration with Albrecht Penck in the years immediately following the Versailles Treaty (see Chapter 2, "German Initiatives").

259 Kleo Pleyer, *Bericht über die im Rahmen der Ostlandtagung des VDA Pfingsten 1935 veranstaltete Tagung des volkswissenschaftlichen Arbeitskreises zu Warnicken an der Samlandküste (11./12. Juni 1935)*, p. 18 (BA R153/94). The report did not list the maps in question.

260 AA DiA-2, Nr. 11, Bd. 4.

261 Memo dated 13 June 1929 (AA Abt. II (Politik 2), Dt.-österreichischer Volksbund - Österreich, Bd. 1).

262 For example, both Penck and Volz had the title of *Geheimrat*.

263 Karl Keller, an official with the Prussian census office (*Preußisches Statistisches Landesamt*) stated in a publication of the *Reichszentrale für Heimatdienst* that the revision of Germany's eastern boundaries constituted an official program of the German government. In support of this claim, he cited statements by Chancellors Luther and Müller, as well as Foreign Minister Stresemann (Keller 1929: 64–5).

264 Stillich (1930) was one of the few people who tried to draw attention to the revisionist mapping of the VDA. However, he failed to recognize that some of the VDA maps he criticized were also disseminated by the Weimar Ministry of Information (*Reichszentrale für Heimatdienst*). As a result, his efforts to prove that the VDA was reactionary and anti-Weimar Republic by citing these maps were not very effective.

265 Memo dated 13 June 1929 (AA Abt. II (Politik 2), Dt.-österreichischer Volksbund - Österreich, Bd. 1).

8 CONCEPTS OF NATIONAL TERRITORY IN THE THIRD REICH

266 *Mitteilungen des Reichsamts für Landesaufnahme* 14 (1938: 334).

267 In violation of the restrictions on arms production in the Treaty of Versailles, Germany pursued the development of its military machinery as early as the 1920s through secret contracts with the Soviet Union.

268 For another version of this map, see Herb (1989: 297).

269 On the propaganda activities of the publisher of this work, *Volk und Reich*, see Chapter 9.

270 Masaryk's map is reprinted in Heiss et al. (1938: 53). Kuffner's map is illustrated in Unser Staat . . . 1938: 278–9). For a translated redrawing of Kuffner's map in the *Zeitschrift für Geopolitik*, see Herb (1989: 296). Several maps from Czech peace conference memoranda were reprinted in publica-

tions of *Volk und Reich*, which specialized in suggestive maps. For example, Loesch (1937: 412–13, 415), Viererbl (1937: 382), and Heiss et al. (1938: 62, 92–3).

271 Maps depicting unemployment and suicide rates appeared in Schumacher (1937: 21, 237) and Heiss et al. (1938: 257, 261), and a map of the distribution of Czech nationalist organizations was included in Schumacher (1937: 189).

272 See correspondence between *Regierungspräsident* Budding (the initiator of the proposal), Albert Brackmann, the *Publikationsstelle-Berlin*, and the Foreign Office (BA R153/708).

273 For examples of such maps, see Kronacher (1938: 226) and Hagemeyer and Leibbrandt (1939: 162).

274 Bolshevism was presented as being part of the long tradition of this eastern threat. See the map in Kronacher (1938: 226).

275 The NSDAP *Reichsparteitag* in 1938 exploited this theme to the maximum with an exhibition entitled "Europe's struggle of destiny in the East" (*Europas Schicksalskampf im Osten*). A book of the materials shown at the exhibition, which contained numerous maps, was published the following year (Hagemeyer and Leibbrandt 1939). Maps of Germany's cultural legacy in the East were already prominent in the Weimar Republic (see Chapter 7, "The Emergence of New Maps").

276 For a reproduction of a similar map in a geography schoolbook, see Heske (1988: 213).

277 See, for example, a 1944 poster by the *Gaupropagandaamt Oberdonau* (BA Plak 3/28/86).

278 This was clearly stated in a guideline of the NSDAP entitled *Richtlinien der parteiamtlichen Prüfungskommission*, dated 4 November 1938. The guideline is appended to a letter from the Justus Perthes Verlag to Hermann Rüdiger, 24 November 1938 (BA R57/751).

279 For example, the map *"Ostland in Not"* in Leers and Frenzel (1934: 33).

280 Köhler (1987: 222). *Dekan, Rheinisch-Westfälische Technische Hochschule Aachen (Fakultät I)* to *Minister für Wissenschaft, Kunst, und Volksbildung*, 8 November 1934, *Anlage 1: Arbeitsbericht von Overbeck*, 15 October 1934 (Forschungsprojekt Hamburg, Archiv).

281 Hermann Aubin to Albert Brackmann, 15 January 1936, 23 May 1938, 1 December 1938 (BA R153/1305). The atlas was finally published in 1938: Walter Geisler, *Oberschlesien Atlas*, Berlin: Volk und Reich, 1938.

282 See for example, *Amtlicher Führer, Deutschland Ausstellung, Berlin 18. Juli–16. August 1936*, p. 142, 146–7.

283 Several geo-organic maps appeared shortly before the war in the 1939 edition of the book *Deutschland und der Korridor* (Heiss et al. 1939), but they were reprints of maps from the 1933 edition.

284 The report of the Westermann representative mentioned that the resolution was proposed by a teacher from Plauen, *Studienrat* Engelmann (WWA). However, the resolution was not listed in the official publication of the *Geographentag* (*Verhandlungen . . .* 1935).

285 *Richtlinien der parteiamtlichen Prüfungskommission*, dated 4 November 1938. The guidelines are appended to a letter from the Justus Perthes Verlag to Hermann Rüdiger, 24 November 1938 (BA R57/751).

286 *Reichsminister für Volksaufklärung und Propaganda* to Justus Perthes Verlag, 25 May 1938, copy of letter (*Abschrift*) in correspondence file, Hermann Rüdiger–Justus Perthes (BA R57/169).

287 For a discussion of the role of race in Nazi politics and propaganda, see Welch (1993: ch. 4, esp. 65–82).

288 *Neue Anordnungen der parteiamtlichen Prüfungskommission, 3068/I7f/Mo.,* 13 January 1939 (BA R57/751).

289 Günther, also known as *Rassengünther,* was a professor at the Institute for Racial Studies, Ethnic Biology, and Rural Sociology (Burleigh 1988: 213, fn. 239).

290 *Reichserziehungsministerium, Erlass E III P. Nr. 23/39,* 17 January 1939 (BA R57/751).

291 Dehmel 11,2: 1933–45, p. 66 (WWA).

292 Examples of such maps can be found in *Meyers Lexikon* (Leipzig: Bibliographisches Institut, 1937, 8th edition, vol. 2, p. 1324), and in Thalheim and Ziegfeld (1936: 338).

293 Lohse worked for the Foreign Office. Ten thousand copies of the 1939 edition were distributed by the Foreign Office abroad. Internal memo of the Foreign Office, 14 July 1939 (AA, Presseabteilung, Handakten Starke, Nr. 60).

294 Confidential report entitled *Besprechung über die Sprachenkarte im Atlas des deutschen Lebensraumes, Geographisches Institut der Universität Berlin, 13. Dezember 1935* (BA R153/1559).

295 Confidential report entitled *Protokoll der Besprechung im Preuß. Geh. Staatsarchiv, Berlin-Dahlem, behufs Zusammenarbeit bei der Karte des deutschen Volksgebiets im Atlas des deutschen Lebensraumes. 29. 2. 1936* (BA R153/1559).

296 The VDA had two noted authorities on German settlement areas: Friedrich Lange and Gottfried Fittbogen. Fittbogen had prepared a publication for the *Reichszentrale für Heimatdienst* (Fittbogen 1928) and Lange made several large-scale maps, such as a map of the distribution of German language in Central Europe and a map of German culture. The DAI in Stuttgart also could boast two cartographic experts on Germandom, O. A. Isbert and Hermann Rüdiger. Isbert not only wrote articles about the methodology of statistics and cartography, but also conducted research on German *Volksboden* in the East (see references in Isbert 1938). Rüdiger, a geographer by training, was the author of the map Rüdiger–Haack, *Das Deutschtum der Erde.* Gotha: Justus Perthes (several editions). He also had first-hand knowledge of the distribution of Germans in the Danube region, where he had conducted field work in 1922 (BA R57/169).

297 Emil Meynen to Albert Brackmann, 4 October 1935; Franz Doubek, *Abschrift* (BA R153/1529).

298 Albert Brackmann to Foreign Office, 30 January 1935 (BA R153/1143).

299 VDA *Bundesleitung* to NODFG, 13 October 1936 (BA R153/1143). On Doubek, see also Burleigh (1988).

300 On the *Vomi,* see Koehl (1957: 37–8; and 1983: 146–7).

301 *Amt für Schulungsbriefe* report entitled *Betrifft: 97.546.000 Deutsche in aller Welt als Zahlenunterlage zum Programm-Punkt 1 des Partei-Programms,* 21 February 1938 (BA R57/751).

302 Hermann Rüdiger (DAI) to *Amt für Schulungsbriefe,* 11 March 1938 (BA R57/751).

303 Although the participants varied somewhat from meeting to meeting, the list of the people and organizations who were consulted included the following:

Benninghaus (*Bund Deutscher Osten*)
Bannführer Dorfmeister (VDA and *Bund Deutscher Osten*)
Franz Doubek (*Publikationsstelle-Berlin* and NODFG)

Dr Essen (Ministry of the Interior)
Walter Geisler (*Geographisches Institut, Universität Aachen*)
Karl Haushofer (*Geographisches Institut, Universität München* and *Deutsche Akademie*).
Horst Hoffmeyer (*Volksdeutsche Mittelstelle* and *Bund Deutscher Osten*)
O. A. Isbert (DAI)
M. Klante (*Volksdeutsche Forschungsgemeinschaften, Geschäftsstelle*)
Kredel (*Bund Deutscher Osten*)
Dr Krieg (Ministry of Propaganda)
Friedrich Lange (VDA)
Henry Lichtner (*Aussenpolitisches Amt der NSDAP, Dienststelle Rosenberg*)
Karl C. von Loesch (*Institut für Grenz- und Auslandsstudien*, Berlin)
Emil Meynen (*Volksdeutsche Forschungsgemeinschaften, Geschäftsstelle*)
Johannes Papritz (*Publikationstelle-Berlin* and NODFG)
Hermann Rüdiger (DAI)
Stäuber (VDA)
Dr Straka (*Alpenländische Forschungsgemeinschaft*)
Dr Wunderlich (DAI)
Arnold H. Ziegfeld (Edwin Runge Verlag)

The alphabetical list was compiled from the attendance records in the reports for the meetings on 22 April 1938, 30 May 1938, and 30 June 1938, as well as from the list of addresses in a letter from the *Volksdeutsche Mittelstelle* dated 15 June 1938, which informed participants about map censorship (BA R153/110 and BA R153/111). [Note: In the meeting reports and the letter, participants were mostly listed by last names. I added first names and other organizational affiliations when known.]

304 Hermann Rüdiger to Perthes Verlag, 2 March 1938 (BA R57/169).
305 Report, *Volksdeutsche Mittelstelle, Besprechung am 30. Mai 1938 über volkspolitische Karten*, p. 4 (BA R153/110).
306 Report of the *Vomi* meeting on 22 April 1938 (BA R153/110).
307 Report of the *Amt für Schulungsbriefe: Betrifft: 97.546.000 Deutsche in aller Welt als Zahlenunterlage zum Programm-Punkt 1 des Partei-Programms*, 21 February 1938 (BA R57/751).
308 Karl von Loesch, *Rasse, Volk, Staat und Raum in der Begriffs- und Wortbildung*. Berlin: Akademie für Deutsches Recht, Ausschuss für Nationalitätenrecht, Unterausschuss für terminologische Angelegenheiten, 1938. Cited after Roessler (1990: 66, 242). The report was labeled "confidential."
309 Karl C. von Loesch, *Bemerkungen zu den Ergebnissen der Besprechung über gross-maßstäbige Volkstumskarten mit Flächenkolorit am 22. 4. 38 in der Volksdeutschen Mittelstelle*, 17 May 1938 (BA R153/110).
310 For example, the German encyclopedia *Meyers Lexikon* included such a map in its 1937 edition. The title of the map was "The distribution of Germanic blood or rather the addition of German blood in Europe." The legend of the map had the following categories: 1) distribution of the German population in Europe; 2) strong Germanic race addition because of the migrations of peoples (*Völkerwanderung*). (*Meyers Lexikon*. Leipzig: Bibliographisches Institut, 1937, 8th edition, vol. 2, p. 1324).
311 See, for example, H. Schrepfer, "Raum, Rasse und Volk," in *Verhandlungen . . .* (1935: 65–84). The only critical remark during the discussion was by Leo Waibel, who had to emigrate a few years later. See also the documentation in Sandner (1990). On the introduction of racial elements in *Geopolitik*, see Bassin (1987b).

312 Report on the *Vomi* meeting of 30 May 1938 (BA R153/110).

313 Walter Geisler, *Erläuternde Bemerkungen zum Entwurf einer Karte des deutschen Volkstums im deutschen Nordosten*, 2 April 1938. Copy of paper appended to a letter from the *Bund Deutscher Osten* to Hermann Rüdiger, 12 April 1938 (BA R57/169).

314 Working paper by O. A. Isbert, *Zur kartographischen Besprechung in der Volksdeutschen Mittelstelle, Montag, d. 30. Mai 1938* (BA R153/110).

315 E.g. "Die steirische Sprachgrenze, 1:150,000," *Deutsches Archiv für Landes- und Volksforschung* 1 (1937).

316 The incomplete references in the working paper for suggestive maps were: "black and white sketch map" in Wutte-Ziegfeld, *Taschenbuch des Grenz- und Auslandsdeutschtums*; and *Volk und Reich*, Beiheft: "Die Südostdeutsche Volksgrenze."

317 Compiled from the reports for the meetings on 22 April 1938, 30 May 1938, and 30 June 1938 (BA R153/110 and BA R153/111).

318 The group decided to depict areas which had experienced a decline in the number of Germans in recent years with so-called "expulsion crosses" (*Verdrängungskreuze*).

319 The following description is based on the records of the meetings on 30 May 1938 (BA R153/110) and 30 June 1938 (BA R153/111).

320 The line of demarcation was the armistice boundary between Polish and German troops which was decreed by the Allied and Associated Powers on 16 February 1919.

321 There was considerable discussion about the corridor region. Some of the participants believed that the boundary of category I corresponded more or less with contemporary political boundaries, while others, notably Walter Geisler, argued that the *Volksboden* reached much further east. The differences in opinion were related to the expulsion of Germans which had taken place in the corridor region after the Treaty of Versailles. The group of participants which wanted to limit the German *Volksboden* to the existing political boundaries thought that the decline in the number of Germans in the corridor region did not allow further-reaching claims, while Geisler and others were convinced that it still was German *Volksboden*. The solution adopted was suggested by Johannes Papritz from the *Publikationsstelle-Berlin*.

322 Meeting notes of 30 May 1938 (BA R153/110).

323 *Richtlinien der parteiamtlichen Prüfungskommission*, dated 4 November 1938. The guidelines are appended to a letter by the Justus Perthes Verlag to Hermann Rüdiger, 24 November 1938 (BA R57/751).

324 AA Abt. Inland IIc, Ki. 4 D 3, *Handwörterbuch des Grenz- und Auslandsdeutschtums*, Band 41/2 (1943–4). The *Handwörterbuch* originally was a project of the *Stiftung für deutsche Volks- und Kulturbodenforschung* (see Chapter 5).

325 Foreign Office report, *Arbeitstagung der Wissenschaftsinstitute und Publikationsstellen in Prag vom 9. bis 10. 3. 1944* (AA Abt. Inland IIc, Ki.1, D2-VFG Bd. 14, folio D 653380).

326 Erwin Winkler, *Nationalitätenkarte der Sudetenländer (Böhmen, Mähren-Schlesien)*, 1:750,000, Karlsbad-Leipzig: Karl H. Frank, 1936 (BA Kart 884/1).

327 "Die Siedlungsgebiete der Deutschen in der Tschechoslowakei," 1:1,125,000, *Deutsches Archiv für Landes- und Volksforschung* (2. Jg. 1938, Heft 2). A copy of the map is preserved at the *Bundesarchiv* (BA Kart 884/3 and 884/11).

328 On the *Forschungsgemeinschaften*, see Burleigh (1988) and Roessler (1990).

329 For example, Winkler (1927). Foreign Office memo, 1929 (AA DiA-2, Volz-Stiftung, Bd. 2).

330 Erwin Winkler, *Gemeindegrenzenkarte der Sudetenländer*, 1:75,000, no date (BA-Kart R113, Rep. 325).

331 Foreign Office memo, 1929 (AA DiA-2, Volz-Stiftung, Bd. 2).

332 *Vertraulicher Bericht über die Tagung des Volkswissenschaftlichen Arbeitskreises im VDA am 5. und 6. Januar 1939 in Bayreuth*, p. 31 (BA R153/96).

333 The *Publikationsstelle-Berlin* was the publishing division of the *Nord- und Ostdeutsche Forschungsgemeinschaft.*

334 *Publikationsstelle-Berlin* to *Bund Deutscher Osten*, 21 July 1939, p. 1 (BA R153/256).

335 *Publikationsstelle-Berlin* to *Bund Deutscher Osten*, 21 July 1939 (BA R153/256).

336 Franz Doubek, *Arbeitsbericht für das Berichtsjahr 1939–40*, p.1 (BA R153/1629).

337 Brackmann (*Publikationsstelle-Berlin*) to *Staatsminister und Preußischer Finanzminister* Dr Popitz, 27 September 1939 (BA R153/257).

338 F. A. Doubek, *Die Bevölkerung Polens. Deutsches Volkstum im Nationalitätengefüge des polnischen Staates*, 1:1,000,000, Berlin: Reichsministerium des Innern, Publikationsstelle-Berlin, 1939 (BA-Kart 778/19 and 878/2).

339 Franz Doubek, *Arbeitsbericht für das Berichtsjahr 1939–40*, p. 2 (BA R153/1629).

340 Franz Doubek, *Arbeitsbericht für das Berichtsjahr 1939–40*, p. 2 (BA R153/1629). Brackmann (*Publikationsstelle-Berlin*) to *Staatsminister und Preußischer Finanzminister* Dr Popitz, 27 September 1939 (BA R153/257).

341 NSDAP *Gauleitung Schlesien, Amt des Gaubeauftragten Schlesien für aussenpolitische Fragen im Stabe des Stellvertreter des Führers* to *Publikationsstelle-Berlin*, 24 June 1939 (BA R153/748).

342 Franz Doubek and Erna Horn, *Die Sprachenverteilung in Westpreußen auf Grund der Erhebungen der Volkszählung im Jahr 1910 über die Muttersprache auf dem flachen Lande. Nach Gemeindeeinheiten dargestellt*, 1:300,000, Berlin: Reichsministerium des Innern, Publikationsstelle, no date (BA-Kart 851/3). Franz Doubek and Erna Horn, *Die Sprachenverteilung in Westpreußen auf Grund der Erhebungen der Volkszählung im Jahr 1910 über die Muttersprache in den Städten, im Flächenverhältnis zur Bevölkerungsdichte des flachen Landes*, 1:300,000, Berlin: Reichsministerium des Innern, Publikationsstelle, no date (BA-Kart 851/2).

343 The map had a scale of 1:1,000,000 (BA-Kart 857/1).

344 *Deutsches Ausland-Institut* to *Oberkommando der Wehrmacht*, 13 January 1939 (BA R57/221).

345 Ministry of the Interior to *Publikationsstelle-Berlin*, 3 July 1939, "Geheim!" (BA R153/257).

346 *Oberkommando des Heeres, 9. Abt. (IV-Mil.Geo), Generalstab des Heeres* to *Publikationsstelle-Berlin*, 28 August 1939 (BA R153/257)

347 This was stated in Franz Doubek, *Arbeitsbericht für das Berichtsjahr 1939–40*, p. 2 (BA R153/1629).

348 *Publikationsstelle-Berlin* to *Studienassessor* Stuber (*Oberkommando der Wehrmacht*), 2 February 1940 (BA R153/740). I have not been able to locate a copy of the map so far.

349 For example, M. Klante (Sammlung Georg Leibbrandt), *Völkerkarte der Sowjetunion (Europäischer Teil). Dargestellt nach der Bevölkerungszählung der UdSSR von 1926 auf der Verwaltungseinteilung nach dem Grossen Weltatlas der UdSSR von 1938*, 1:5,000,000, Berlin: Reichsamt für Landesaufnahme, 1941 (BA Kart 873/10). Sammlung Georg Leibbrandt, *Die deutschen Siedlungen in der Sowjetunion (5 Teile)*, 1:1,000,000 and 1:300,000, Berlin: Reichsamt für Landesaufnahme, 1941 (BA-Kart 873/4 to 873/8). See also the lists of maps appended to Franz Doubek, *Arbeitsbericht für das Berichtsjahr 1939–40*, which cover the cartographic activities of the *Publikationsstelle-Berlin* between 1939 and December 1942 (BA R153/1629) and the list of maps appended to the letter from the *Deutsches*

Ausland-Institut to the High Command of the Army, 13 January 1939 (BA R57/221). A number of maps are also mentioned in the report of the *Südostdeutsche Forschungsgemeinschaft* appended to the letter from the *Geschäftsstelle der Volksdeutschen Forschungsgemeinschaften* to the Foreign Office, 31 March 1944 (AA Abt. Inland IIc, VFG/Deutschtum 2, Bd. 14).

350 Foreign Office report, *Arbeitstagung der Wissenschaftsinstitute und Publikationsstellen in Prag vom 9. bis 10. 3. 1944* (AA Abt. Inland IIc, VFG/Deutschtum 2, Bd. 14, folio D 653385–6).

351 Interior Ministry to *Volksdeutsche Mittelstelle*, 11 November 1939 (BA R153/258).

352 *Zusammenstellung der vom 1.4.1940 bis 1.4.1941 bei uns hergestellten wichtigsten Karten*, p. 1 (BA R153/1629). On the SS-*Reichskommissariat für die Festigung des deutschen Volkstums*, see Koehl (1957).

353 Franz Doubek, *Arbeitsbericht für das Berichtsjahr 1939–40*, p. 2 (BA R153/1629).

354 Interior Ministry to Foreign Office, High Command of the Army, *Volksdeutsche Mittelstelle*, and *Publikationsstelle-Berlin*, 16 January 1940 (BA R153/740).

355 Franz Doubek, *Bericht über meine dienstliche und wissenschaftliche Tätigkeit im Kalenderjahr 1941*, p. 2 (BA R153/1629).

356 The symbols had the following range: less than 10; 10; 100; 1,000; 10,000; 100,000 and more. *Die Verteilung der Juden im nordwestlichen europäischen Russland, 1:1,500,000 (Punktkarte)*, Berlin: Reichsministerium des Innern, Publikationsstelle, 1942 (BA-Kart 873/1).

357 The most comprehensive listing of these institutes can be garnered from the report on a telephone conversation between a member of the Foreign Office and SS-*Hauptsturmführer* Krallert, dated 2 March 1944, and from the meeting schedule of the "Prager Arbeitstagung der wissenschaftlichen Einrichtungen zur Landes- und Volksforschung." Internal memo of the Foreign Office, dated 15 March 1944 (AA Abt. Inland IIc, VFG/Deutschtum 2, Bd. 14, folios D 653298–9 and D 653304–5).

358 For a description of raided materials used by the *Publikationsstelle-Berlin*, see Burleigh (1988: 227–35).

359 *Volkstumskarte der Krim*, 1:400,000, Wien: Publikationsstelle in collaboration with Sammlung Georg Leibbrandt (BA-Kart 858/11).

360 Krallert also participated in the raids of the SS-*Sonderkommando* "Gruppe-Künsberg," which enabled him to get hold of the most accurate data on the distribution of ethnic groups in Russia. Internal memo of the Foreign Office, *Betrifft: Volkstumskarte 1:1,000,000 der Publikationsstelle Wien*, 5 October 1943 (AA Abt. Inland IIc, Ki.1, D2-VFG Bd. 14). On the SS-*Sonderkommando* "Gruppe-Künsberg" activities, see also Burleigh (1988: 227–35).

361 Krallert (*Publikationsstelle-Wien*) to Foreign Office, 9 March 1942 (AA Abt. Inland IIg, 288, *Volkstum und Grenzfragen*).

362 Internal memo of the *Publikationsstelle-Berlin*, 18 March 1944 (BA R153/745).

363 The greater accuracy of Penck's dot-map version and its successful application by the *Publikationsstelle-Wien* was mentioned in an internal memo of the Foreign Office, dated 18 October 1943 (AA Abt. Inland IIc, Ki.1, D2-VFG Bd. 14).

364 *Lettland. Volkliche Gliederung in den politischen Gemeinden*, 1:200,000, Sicherheitsdienst, Hauptamt II (NO), no date (BA-Kart 877/5).

365 Internal memo of the *Publikationsstelle-Berlin*, 31 August 1939 (BA R153/277).

366 *Bericht über die Tätigkeit der Kartenabteilung der Publikationsstelle (1.1.1942–31.12. 1942)* (BA R153/1629).

367 *Publikationstelle-Berlin* to Interior Ministry, 9 May 1939; President of the

Reichsamt für Landesaufnahme to *Publikationsstelle-Berlin*, 26 May 1939 (BA R153/ 136).

368 An important part of the cartographic production of the *Publikationsstelle-Berlin* was the preparation of maps depicting administrative boundaries, which were distributed to various government offices for use as base maps.

369 *Publikationsstelle-Berlin* to Ministry of the Interior, 4 October 1939 (BA R153/ 280).

370 *Erlass des Reichsamts für Landesaufnahme Kart 5200 d.* The decree is mentioned in *Richtlinien für die Darstellung politischer Verhältnisse auf Landkarten,* 25 February 1943, p. 7 (AA Abt. Inland IIg, 288, *Volkstum und Grenzfragen*).

371 *Presserundschreiben des Reichspropagandaamtes Berlin, Nr. II/382/40, geheim* (BA R55/1386, folio 79).

372 Luther to Dr Krümmer, 20 July 1942 (AA R27634). Apparently, the project was not just an outgrowth of the German belief in a speedy victory, but also a response to foreign activities: Hitler had expressed great discontent that countries in Southeastern Europe were presenting him with atlases on a regular basis, which showed their territorial development, while Germany had no reliable atlas of this kind.

373 AA Abt. Inland IIc, *Historisch-geographischer Atlas von Europa*, Deutschtum–2, Bd. 2.

374 PRO FO1031/140, p. 218.

375 Communication by Dr Woldan through the mediation of Dr Lazar, Vienna.

9 MAPS AND NAZI PROPAGANDA

376 For example, see Welch (1993), Kershaw (1983b), Bessel (1980), and the contributions in Welch (1983).

377 For a thorough discussion of the restructuring of the means of communication under the Nazis, see Welch (1993: ch. 3).

378 Erwin Schockel, *Das politische Plakat. Eine psychologische Betrachtung,* Munich, 1939: 122. Cited after Paul (1990: 150).

379 Hitler stated in *Mein Kampf* that he was also influenced by "Marxist-Socialist" propaganda (Welch 1993: 10).

380 Berlin: Verlag für Literatur und Politik. Reprinted in the GDR in 1980.

381 The meeting did not produce any tangible results, but it led to the expression "Harzburg Front" to describe their shared interests (Koehl 1983: 50). See also Mosse (1981: 223–5, 250–1).

382 His full name was Rupert Franz Schumacher Ritter von Tännengau (BDC, personal file R. von Schumacher).

383 Schumacher's rejection of italics appears at first sight to be in contradiction to Karl Springenschmid's use of *Sütterlin* (see Chapter 6). However, Springen-schmid's maps were sketches intended for use in schools, where *Sütterlin* was introduced as the official German handwriting style between 1935 and 1941.

384 Schumacher (1935). The catalog of symbols is reproduced in a translated version in Whittlesey et al. (1942: 130–9).

385 The evaluation from July 1940 addressed to the personal staff of the *Reichs-führer* SS stated that Schumacher's work was generally solid, but constituted neither a "scientific" work, nor a "readable popular" treatment (BDC, personal file R. von Schumacher). The complete reference for the book is Rupert von Schumacher, *Des Reiches Hofzaun. Geschichte der deutschen Militär-grenze.* Darmstadt: L. Kichler, 1940.

386 SS-*Hauptsturmführer* A. Meine to Dr Korherr (*Statistisch-Wissenschaftliches Institut des Reichsführer* SS), 16 August 1944 (BDC, personal file R. von Schumacher).

387 Schumacher joined the Austrian National Socialist students association in 1929 and the Austrian Nazi party on 29 April 1930. He also was a member of the personal staff (*Referent*) of *Reichsbauernführer* Darrée (ca. 1940) and somehow involved with the *Stabshauptamt des Reichskommissars für die Festigung des Deutschen Volkstums*. More detailed information is not available (BDC, personal file R. von Schumacher).

388 *Ministerium für Wissenschaft, Erziehung und Volksbildung*, Erlass E IIa 5068, 8 March 1940 (BA R21/74, folios 179–90).

389 Standardization of suggestive cartography was proposed to the Propaganda Ministry as early as 1933 by Arnold Ziegfeld (see Chapter 9, "Organization of Production".

390 Meynen's book was published by the *Zentralkommission für wissenschaftliche Landeskunde von Deutschland*, a commission of the Association of German Geographers (*Geographentag*) which was headed by Friedrich Metz, an outspoken *völkisch* nationalist.

391 For examples of such attacks, see Chapter 9, "Regulation and Control".

392 The poster is preserved at the *Bundesarchiv* (BA Plak 3/3/53).

393 The committee included members of the Propaganda Ministry, the Foreign Office, the Ministry of Education, *Reichsschrifttumskammer, Geheime Staatspolizei, SS-Sicherheitshauptamt*, and the *Parteiamtliche Prüfungskommission*. (Minutes of the committee meetings, 18 June 1936 and 4 July 1936; BDC, personal file Emil Meynen.)

394 BDC, personal file A. H. Ziegfeld. Ziegfeld's 1935 article makes a reference to this proposal (p. 246), but lists 1932 as the year, which is obviously wrong.

395 Undated report on Ziegfeld's achievements in the Propaganda Ministry and fragment of a paper on the *Kartenstelle* (BDC, personal file A. H. Ziegfeld).

396 BDC, personal file A. H. Ziegfeld.

397 BDC, personal file A. H. Ziegfeld.

398 Lehmann (*Bibliographisches Institut*) to Ziegfeld (Propaganda Ministry), 2 March 1940 (AA, Presseabteilung, P. 16, Bildmaterial, Bd. 3). Propaganda Ministry to Foreign Office, 20 December 1939 (AA, Presseabteilung, P. 16, Bildmaterial, Bd. 2). An undated report on Ziegfeld's achievements in the Propaganda Ministry, which is included in his personal file at the BDC, mentions that the map service resulted in the production of more than 3,000 maps. But this figure might be exaggerated since the report seems to have been authored by Ziegfeld. Ziegfeld had submitted a report on his activities in the Propaganda Ministry to the personnel division on 4 May 1942, but it was sent back as being unsolicited (*Personalabteilung* to Ziegfeld, 13 May 1942; BDC, personal file A. H. Ziegfeld). On 1 June, one of Ziegfeld's superiors submitted a report on Ziegfeld, which seems to have been the same report judging from a handwritten note on the accompanying letter (Fitz-Rudolf, *England-Ausschuss*, to Flügel, *Personalabteilung*, 1 June 1942; BDC, personal file A. H. Ziegfeld).

399 Report of the *Amt für Schulungsbriefe: Betrifft: 97.546.000 Deutsche in aller Welt als Zahlenunterlage zum Programm-Punkt 1 des Partei-Programms*, 21 February 1938 (BA R57/751). These numbers appeared on the map and in tables of the April 1938 *Schulungsbrief*. The map was confiscated shortly after its publication because it depicted South Tyrol as German *Volksboden*, which was prohibited in 1938. A copy of the map is kept at the Bundesarchiv (BA-Kart 861/1). It is also reproduced rather poorly in Grimm (1939). For a discussion of the censorship incident, see Chapter 9, "Regulation and Control."

400 BA R55/717, folio 93–4. One BDO propaganda postcard even used a thumb-wheel to show the territorial losses in Upper Silesia. It is preserved at the

Bundesarchiv (BA R153/72) and also reproduced in black and white in Burleigh (1988: 74).

401 BA R55/717, folio 83, 90.

402 *Vorstand der Stiftung Volk und Reich* to Foreign Office, 23 March 1932 (AA DiA-2, Volk und Reich, Bd. 1).

403 Friedrich Heiss (*Volk und Reich*) to Foreign Office, 21 July 1933 (AA, DiA-2, Volk und Reich, Bd. 1).

404 AA Presseabteilung, Handakten Starke, Nr. 60, 61.

405 AA Presseabteilung, Az.: Hefte (Paket 482), Monatshefte Volk und Reich, Allgemeiner Schriftwechsel. Some of the maps had been used in other publications before. For an example of a black and white map included in the issue, see Herb (1989: fig. 5).

406 *Volk und Reich* to *Informationsstelle* of the Foreign Office, no date (AA Presseabteilung, Handakten Starke, Nr. 58, 59). The map was entitled "Then and Now! 1914 and 1939" (*Facts in Review*, vol. 1, no. 17, 8 December 1939: 1). It is reprinted in Monmonier (1991: 100).

407 The list of influential persons included, among others, professors, teachers, physicians, government employees, executives of industry and commerce, and military personnel. *Deutsche Gesandschaft Stockholm* to Foreign Office, 19 December 1939 (AA Abt. Inland I, Propaganda, Bd. 1).

408 Waldemar Wucher (*Volk und Reich*) to Foreign Office, 2 February 1940, with attached copy of the title page of *Facts in Review* (AA Presseabteilung, P16c-Propaganda im Inland, Bd. 2).

409 Correspondence with Wolfgang Höpker, Bonn, who was one of the members of the editorial board of the atlas. Many of the atlas maps were badly designed, which makes it unlikely that *Volk und Reich* produced them. See also Herb (1989).

410 A. Ziegfeld to Foreign Office, no date, but according to the stamp at head of letter it was received at Foreign Office on 8 December 1933; *Volk und Reich* to Foreign Office, 2 February 1935 (AA DiA-2, Volk und Reich, Bd. 1).

411 *Volk und Reich* to Foreign Office, 14 November 1933 (AA DiA-2, Volk und Reich, Bd. 1).

412 AA DiA-2, *Institut für Grenz- und Auslandsstudien*, Bd. 1.

413 H. Rüdiger to Justus Perthes Verlag, 5 July 1939 (BA R57/169).

414 AA Presseabteilung, Handakten Starke, Nr. 60, 61.

415 The maps for the indoctrination (*Schulung*) of the Hitler Youth in 1941 included the following: 1) four maps of war campaigns (Poland, Norway, etc.) with a circulation of 40,000 copies; 2) a map entitled "Die deutschen Lande" (10,100 copies); 3) a twelve-map series entitled, "Der Kampf um das Reich" (31,000 copies); 4) a four-map series, "Das Werden des Großdeutschen Reiches" (41,000 copies). AA Presseabteilung, Handakten Starke, Nr. 61, 62.

416 AA Presseabteilung, Handakten Starke, Nr. 61, 62.

417 Heiss, the director of *Volk und Reich*, was an *SS-Obersturmbannführer im Persönlichen Stab des Reichsführer-SS* in 1940. This title appeared on a letter dated 22 October 1940 (AA Presseabteilung, Handakten Starke, Nr. 61, 62).

418 AA Presseabteilung, Handakten Starke, Nr. 61, 62.

419 Heiss to Lohse (Foreign Office), 8 August 1939 (AA Presseabteilung, Handakten Starke, Nr. 60).

420 There were other *Publikationsstellen*, such as the *Publikationsstelle-Wien* and *Publikationsstelle-Ost*, but they do not appear to have been as involved in the production of suggestive maps as the *Publikationsstelle-Berlin*, which was

conveniently located. Berlin was the seat not only of the government, but also of central party offices and *völkisch* organizations.

421 The propaganda division of the German High Command also sought the expertise of the *Puste*. In April 1939, it asked the *Puste* for two of its ethnographic maps, which the propaganda division believed "could be of great interest" for its work. *Oberkommando der Wehrmacht (W Pr IV)* to *Publikationsstelle-Berlin*, 22 April 1939 (BA R153/740).

422 "Protokoll über die Besprechung am 1. 11. 1933 im Osteuropa Institut" (BA R153/233); Johannes Papritz, "Ostdeutsche Vergangenheit. Ein Führer durch die geschichtliche Abteilung der Ausstellung 'Der Osten – das deutsche Schicksalsland,' veranstaltet vom B.D.O." *Ostland* 51, 52 (1933), Sonderdruck (BA R153/283); *Puste* translation, dated 20 December 1933, of the 14 December 1933 edition of *Gazeta Polska* (BA R153/233).

423 Aubin was a member of the ruling committee of the *Nord- und Ostdeutsche Forschungsgemeinschaft* (NODFG) and Kuhn's research was continuously supported by the NODFG (Burleigh 1988: 72, 103). Doubek was in charge of the cartographic production of the *Puste* and NODFG from 15 October 1936 (letter VDA to NODFG, 13 October 1936; BA R153/1143).

424 Correspondence between BDO and Doubek (*Publikationsstelle-Berlin*), 11 March 1937, 20 March 1937, 30 March 1937, 8 November 1937, 1 December 1937, 7 December 1937, 8 December 1937 (BA R153/748).

425 *Publikationsstelle-Berlin* to *Amt für Schriftumspflege*, 4 July 1938; *Publikationsstelle-Berlin* to *Institut für Allgemeine Wehrlehre*, 13 July 1938 (BA R153/115). Photographs of the final drafts of the maps are included in BA R153/1070.

426 Circular letter by Niedermayer to all scientific collaborators on the 1938 party exhibit, 22 July 1938 (BA R153/115).

427 NSDAP *Reichsleitung* to Papritz (*Publikationsstelle-Berlin*) 15 October 1938 (BA R153/1070).

428 Circular letter by *Institut für Allgemeine Wehrlehre* to all scientific collaborators on the 1938 party exhibit, 4 October 1938 (BA R153/1070).

429 *Publikationsstelle-Berlin* to *Institut für Allgemeine Wehrlehre*, 10 November 1938 (BA R153/1070).

430 *Publikationsstelle-Berlin* translation of the *Briva Zeme* article of 10 September 1938 in "Lettische Presseauszüge (Nr. 88)" (BA R153/1070).

431 Regulation and censorship in this context only refers to the political content and design of maps; censorship for reasons of military security, which restricted cartographic depictions of militarily sensitive targets, such as factories, is excluded from the discussion.

432 For example, the process of cartographic Germanization in ethnographic maps in the Weimar Republic that was described in Chapter 7 ("Changes in Map Design, School Atlases") continued after 1933. In the Diercke atlas, the Germanization mission was completed by the second printing of the 1939 edition: the entire area of the map *Deutsches Reich, Völker- und Sprachenkarte* which previously had been depicted as mixed German/Slavic was now depicted as purely German. Thus, in the year in which Hitler invaded Poland all lost eastern territory (except for the district of Poznan) had been cartographically reclaimed. These changes do not appear to have been the result of Nazi censorship since there are no records for specific decrees or official requests to change the design of ethnographic maps in the WWA.

433 A collection of quotes can be found in Sandner (1990).

434 Paul Diercke, *Diercke Atlas Oberstufe, Neubearbeitung 1933* (WWA). An indication of Diercke's voluntary adaptation to the new ideology is the extension of the areal coverage of the ethnographic map *Deutsches Reich, Völker- und*

Sprachenkarte in the 1933 edition of his atlas. The southern boundary of the map was extended to show the German territory of Austria and parts of Switzerland, the western to allow for the depiction of the entire Dutch component of the "Germanic territory" category.

435 For example, a travel advertisement which read "Get to know Germany" (*Lernt Deutschland kennen*) and displayed the political boundaries of the German Empire was reprinted with the commentary "This makes us angry!! Germany?? well, well, it is surely larger" (*Das ärgert uns!! Deutschland?? na, na, das ist doch grösser* (*Rolandblätter*, vol. 9, November 1935: 341).

436 Criticism was mainly directed at the following maps: H. Wehrli, *Neue Völker- und Sprachenkarte von Europa*, Bern: Kümmerly & Frey, 1934 (1:10,000,000); Charles Burky, *Carte éthnique et linguistique de l'Europe centrale*, Bern: Kümmerly & Frey, 1938 (1:2,000,000).

437 Metz (1934). The map was cited as *Verkehrskarte von Deutschland*, 1:1,000,000. Berlin: Gea Verlag, no date.

438 Ministry of the Interior, Runderlass 29 October 1936 (VI A 14109/6460) and Runderlass 28 April 1937 (VI A 4397/6460). See *Mitteilungen des Reichsamts für Landesaufnahme* 14 (1938): 123–9.

439 On the research goals of the *Publikationsstelle-Berlin*, see Burleigh (1988).

440 Gerhard Fischer, "Kartenserie zur Geschichte des polnischen Volkes. *Schulungsbrief des Bundes Deutscher Osten*, Nr. 28, October 1936. A copy of the issue is preserved in BA R153/386.

441 *Publikationsstelle-Berlin* to Professor Oberländer (*BDO*), 24 November 1936 (BA R153/386).

442 *Stellungnahme zum 28. Schulungsbrief des Bundes Deutscher Osten vom Oktober 1936, Kartenserie zur Geschichte des polnischen Volkes von G. Fischer* (BA R153/386).

443 The map, which was entitled "The position of Czechoslovakia in relation to the large regions of Europe," was included in an article by O. Schäfer (1937: 422).

444 *Publikationsstelle-Berlin* to *Schriftleitung der Zeitschrift für Geopolitik*, 30 July 1937; Ministry of the Interior to *Schriftleitung der Zeitschrift für Geopolitik*, 6 September 1937 (BA R153/623).

445 Ministry of the Interior to *Publikationsstelle-Berlin*, 14 August 1939; *Publikationsstelle-Berlin* to Ministry of the Interior, 23 August 1939. The only references to the maps were "BDO, Landesgruppe Schlesien (Dr. Rogmann, Breslau), 1. Die Nationalitäten im Ostraum, 2. Polen – ein Nationalitätenstaat" (BA R153/740).

446 Internal memo of the Foreign Office, "DVIII 281/42g," dated 1 August 1942; *Publikationsstelle-Innsbruck*, "Betreff: Übersichtskarte der Binnenschiffahrtsstrassen für Mitteleuropa," 23 July 1942 (AA, Pol. XVII Geheim, Bd. 3).

447 Thirteenth edition, Leipzig: List & von Bressendorf, 1938.

448 City archive Elbing to A. Brackmann, 21 May 1938 (BA R153/379).

449 *Publikationsstelle-Berlin* to Ministry of the Interior and others, 1 June 1938 (BA R153/379).

450 *Publikationsstelle-Berlin* to city archive Elbing, 1 June 1938 (BA R153/379).

451 City archive Elbing to *Publikationsstelle-Berlin*, 22 October 1938 and 9 November 1938 (BA R153/373).

452 Hitler's order is mentioned in a letter by the NSDAP to the Foreign Office, 15 August 1941 (AA, Pol. XVII Geheim, Bd. 3). The color map is preserved at the Bundesarchiv (BA Kart 861/1) and reproduced rather poorly in Grimm (1939: 12).

453 Perthes Verlag to H. Rüdiger, 12 May 1938; H. Rüdiger to Justus Perthes Verlag, 20 May 1938 (BA R57/169).

454 The DKG, a private association, was founded in 1937. Carl Wagner, the president of the DKG, claimed that the organization's function was to act as an "official" liaison office between government institutions and private cartographic organizations and firms (Wagner 1938: 343–4). However, there are no records of activities which went beyond the dissemination of general information.

455 H. Rüdiger to R. v. Klebelsberg, 5 September 1938; F. Metz to H. Rüdiger, 6 October 1938; H. Rüdiger to F. Metz, 17 October 1938; *"Abschrift aus dem Brief d. Herrn Carl Wagner (Deutsche Kartographische Gesellschaft e. V.), Leipzig, vom 8. Juni 1938"* (BA R57/169).

456 F. Metz to H. Rüdiger, 6 October 1938 (BA R57/169).

457 See, for example, the 1938 edition of *Sydow Wagners methodischer Schulatlas* (Troll 1947: fn. 10). The Westermann Verlag even had to issue a second printing of its 1938 Diercke school atlas edition to comply with the new requirements (WWA).

458 VDA to Foreign Office, 10 May 1938; Memo of the Foreign Office, 17 May 1938 (AA DiA-2, VDA, Bd. 17).

459 The maps were: Haack-Hertzberg, *Die Völker Mitteleuropas*, 1:1,000,000, Gotha: Justus Perthes Verlag, 1917 with additions in 1920 (Physikalischer Handatlas, Nr. 7); Haack-Hertzberg, *Die Völker Europas*, 1:3,000,000, Gotha: Justus Perthes Verlag, no date (Physikalischer Handatlas, Nr. VII, 7); Arthur Haberlandt, *Die Völker Europas nach Sprache und Volksdichte*, 1:3,000,000, Wien: G. Freytag & Bernd A.G., 1927. *Publikationsstelle-Berlin* to Ministry of the Interior, 3 May 1938, with copies to the Propaganda Ministry, Foreign Office, Education Ministry, *Vomi*, BDO; *Publikationsstelle-Berlin* to Education Ministry, 23 August 1938 (BA R153/623).

460 Propaganda Ministry to Ministry of the Interior, 31 May 1938; Education Ministry to Ministry of the Interior, 11 July 1938 (BA R153/623).

461 This is mentioned in *Vomi* to participants of map meetings, 15 June 1938 (BA R153/110). The only references listed were: 1) "Mitteleuropa – Völker- und Sprachenkarte (Geographische Anstalt H. Wagner & E. Debes, Leipzig)," which was included in the school book by Stephan-Wilke, *Unsere Muttersprache*, Halle/S.: Verlag Hermann Schroedel, Pädagogischer Verlag; 2) The map on p. 135 of W. E. Peuckert, *Schlesische Volkskunde*, Leipzig: Verlag Quelle & Meyer, 1928; 3) The map "Sprachenkarte von Schlesien," on p. 137 of Stanislaw Wasilewski, *Im Oppelner Schlesien*, Kattowitz, no publisher, no date; 4) A variety of travel brochures dealing with the Sorbian question.

462 *Vomi* to participants of map meetings, 15 June 1938 (BA R153/110); *Vomi* report: "Volksdeutsche Mittelstelle, Besprechung am 30. Juni 1938 über volkspolitische Karten" (BA R153/111).

463 The decree is mentioned in a letter by the BDO *Bundesleitung* to the *Publikationsstelle-Berlin*, 23 August 1938 (BA R153/748).

464 For areas outside its expertise, namely the Southeast, the *Publikationsstelle-Berlin* suggested the cooperation of the *Südostdeutsche Forschungsgemeinschaft* in Vienna. There are no records that cooperation with the *Südostdeutsche Forschungsgemeinschaft* actually took place. BDO *Bundesleitung* to *Publikationsstelle-Berlin* 23 August 1938; *Publikationsstelle-Berlin* to BDO, 29 August 1938; BDO to *Publikationsstelle-Berlin*, 1 September 1938 (BA R153/748).

465 Perthes Verlag to H. Rüdiger, 2 March 1939 (BA R57/169).

466 H. Rüdiger to Perthes Verlag, 6 March 1939 (BA R57/169).

467 Decree by *Reichs- und Preußischer Minister für Wissenschaft, Erziehung und Volksbildung*, E IIa 753/37, E III, EIV, 6 April 1937 (BA R21/73, folio 233).

468 Ministry of the Interior to *Publikationsstelle-Berlin*, 14 August 1939 (BA R153/740).
469 Internal report of the Propaganda Ministry, 17 December 1941 (BA R55/409, folio 152).
470 *Lange-Diercke-Schulatlas*, Neubearbeitung, 721. Auflage; *Westermanns neuer Schulatlas*, begründet von Adolf Liebers, 98. Auflage. The atlases were the only two out of a total of seven which were not approved in the Education Ministry decree EIIa 5068, 8 March 1938 (BA R21/74).
471 Copies of the evaluations by the *NS-Lehrerbund*, Bayreuth, are included in Dehmel 11,2: 1933–45, between pp. 27 and 28 (WWA).
472 Westermann Verlag to *Zentralinstitut*, 21 May 1941 (WWA, Schulverlag IV). The only reference to the map was Berendt, *Westermanns Weltkriegskarte*, 2. Auflage.
473 Division of Map Intelligence and Cartography, 1947, p. 19 (AGS).
474 *Presserundschreiben des Reichspropagandaamtes Berlin betreffend Anweisungen für die Gestaltung der Presse, Nr. II/32/40 (25. 1. 40), Geheim* (BA R55/1386, folio 11).
475 Dehmel 11,2: 1933–45, pp. 63–8 (WWA).
476 Justus Perthes Verlag to Westermann Verlag, 20 March 1941; Justus Perthes Verlag to *Reichsstelle für das Schul- und Unterrichtsschrifttum*, 3 March 1942 (WWA).
477 Perthes Verlag to H. Rüdiger, 2 March 1939; H. Rüdiger to Perthes Verlag, 6 March 1939 (BA R57/169).
478 Perthes Verlag to H. Rüdiger, 2 March 1939 (BA R57/169).
479 Perthes Verlag to H. Rüdiger, 21 April 1939 and 26 June 1939 (BA R57/169).
480 Dehmel 11,2: 1933–45, p. 21 (WWA).
481 Education Ministry, decree EIIa 5068, 8 March 1940 (BA R21/74, folios 179–90). See also Dehmel, 11,2: 1933–45, pp. 21–7 (WWA).
482 Dehmel 11,2: 1933–45, p. 25 (WWA).
483 Dehmel 11,2: 1933–45, pp. 25–7 (WWA).
484 *Meldungen aus dem Reich*, 16 February 1940 (BA R58/148, folio 95).
485 Dr Lauda, "Vermerk über die im Ref. Pol. XII GD am Donnerstag, den 22. Juli 1943 stattgefundene Sitzung über 'Vorzensur für Kartenveröffentlichungen in der Presse.'" (AA Abt. Inland IIc, Ki.1 D2-VFG, Bd. 14).
486 Foreign Office to Propaganda Ministry and others, 26 August 1943 (AA Inland IIc, Ki.1, D2-VFG, Hist. geogr. Atlas).
487 The continued use of suggestive maps in schools under the Nazis is revealed in a photograph in *Illustrierter Beobachter* (1933: 1293) which shows a pupil drawing an outline map of the eastern boundaries of Germany during a geography lesson. The sketch indicates the loss of the corridor region and the close proximity of Berlin to the new Polish–German boundary. See also Lissmann (1989: 246), who mentions that pupils in the early years of the Nazi regime were asked to draw a map of the territorial losses stipulated in the Treaty of Versailles in their notebooks.
488 Circular letter by Kleo Pleyer to the members of the *Volkswissenschaftlicher Arbeitskreis im VDA*, 16 November 1936 (BA R153/94).
489 For examples, see *Rolandblätter*, vol. 9, May 1935: 152 (answers to Czechoslovak quiz); and *Rolandblätter*, vol. 9, October 1935: 307 (outline map and questions for Alsace-Lorraine).
490 Welch (1993: 15) states that the emphasis of Nazi propaganda on emotions, which was already outlined in *Mein Kampf*, was carried out with "a ruthless consistency."
491 For an example of an allegorical map distributed as a postcard, which depicts

the impact of the new Polish harbor Gdingen (Gdynia) on Danzig with a snake, see Herb (1993: fig. 9.13). The map is also reproduced in Mehmel (1995: fig. 3). Stretching from its tail in the coalfields of Upper Silesia to its open fanged mouth in Gdynia, the snake is poised to swallow the German port. The title of the postcard map reads: "Poland's corridor harbor Gdingen: the strangler of Danzig."

492 A brochure describing the shrines is included in BA R57/674.

493 A photo of the map appeared in *Illustrierter Beobachter* (1933: Sondernummer).

494 A brochure describing the play is included in AA DiA-2, VDA, Bd. 9.

495 The map is described in the *Publikationsstelle-Berlin* translation of the *Briva Zeme* article of 10 September 1938 in "Lettische Presseauszüge (Nr. 88)" (BA R153/1070). The use of light bulbs on maps was not an invention of Nazi Germany. The Bolsheviks used a map with light bulbs in 1918 to display their plans for the electrification of Russia. The power requirements for the map were apparently so great that the electric lights in the offices of the Kremlin had to be dimmed. The Bolshevik map is described in Alexei Tolstoy, *Ordeal*, vol. 3, Moscow: Foreign Languages Publishing House, 1953: 511–13).

496 The films are preserved at the *Bundesarchiv* (BA Film 760 and BA Film 912).

497 BA Film 912.

498 *Reichspropagandaamt Oberdonau* to Propaganda Ministry, 16 October 1940 (BA R55/725).

CONCLUSION

499 It is necessary to point out that the map in figure 10.1 is highly simplistic. The category "Nationalist demands for a 'Greater Germany'" is too limited. Even if the phrase is interpreted as the bare minimum of nationalist demands, it omits Alsace-Lorraine and the Memel territory, areas customarily claimed by German nationalists.

500 In light of the present research, the argument by Rich (1973: xi) that the variety of claims in Nazi period maps diluted their effect appears inaccurate.

501 Article 116 of the Basic Law and the text of the London Protocol are reprinted in Blumenwitz (1989: 73–4, 168). See also the discussion in Blumenwitz (1989: 28, 32, 37).

502 Associations of Germans who were expelled from the German territories in the East after Germany was defeated.

503 For the nationalist viewpoint on the subject, see Blumenwitz (1980).

504 On the rise of the *Republikaner*, see Leggewie (1990).

505 The only difference is a slight change in the wording of the legend: in the *Republikaner* map the title of the category "German border lands in jeopardy" from the original map has been changed to "areas ceded in the Treaties of Versailles and St Germain."

PRIMARY SOURCES

A. ARCHIVAL SOURCES

American Geographical Society Collection, Milwaukee (AGS)

Division of Map Intelligence and Cartography, "German Cartographic and Map Collecting Agencies: Controling Laws and Regulatory Statutes." Department of State, Intelligence and Coordination Division, Office of Intelligence Coordination: Intelligence Research Report OCL–2996.1, 5 February 1947.

Bundesarchiv, Koblenz (BA)

Filmarchiv
  760, 912.

Kartensammlung
  DAI-Kart 42, 44, 45, 47, 48, 49, 55, 73–82, 105, 112–16, 130, 161–76, 187, 200–8, 214, 226, 227, 234–40, 255, 261, 265, 276, 278, 285, 312, 324, 326, 328–41, 346, 352, 373, 375, 377–91, 408, 410, 414, 424, 435, 439–45, 450, 451, 458, 459, 464, 468, 469, 502–9, 516, 521, 525, 527, 538, 543, 581, 590–6, 598–602, 624, 627, 630, 631, 643, 645, 648, 657, 690–6, 700, 704, 708, 709, 715, 717, 725, 736, 749–52, 756, 757, 760–2, 766, 784, 789, 794, 779–806.

[NOTE: DAI-Kart is a collection of 806 wall maps which belonged to the Deutsches Ausland-Institut. It is probably the most complete wall-map collection relating to German issues available. When I visited the Bundesarchiv in 1989 and 1990, it was not yet cataloged. Dr Hofmann of the Bundesarchiv kindly allowed me to survey the maps and to create a provisional list. The numbers above refer to this list.]

  Kart 778/19, 784/3851/2, 851/3, 856/2, 856/4, 856/5, 857/1, 857/2, 857/6, 857/10, 857/12, 857/14, 858/3–4, 858/11, 858/12, 858/14–16, 861/1, 861/5, 873/1–8, 873/10, 875/1–3, 877/1, 877/2, 877/3, 877/5, 877/8–10, 878/2–5, 881/2, 884/1–4, 884/6–8, 884/10–11.
  Kart R113, Rep. 325

NS 26 (Hauptarchiv)
  266, 1395.

Postersammlung
  Plak 2/7/73, 3/28/86, 3/3/53.

R 21 (Reichsministerium für Wissenschaft, Erziehung und Volksbildung)
73, 74, 75, 83, 84, 85, 503.

R 43 (Reichskanzlei)
I: 518, 360, 361, 1800.
II: 133.

R 55 (Reichsministerium für Volksaufklärung und Propaganda)
409, 464, 717, 725, 1299, 1345, 1366, 1386, 1387, 1440.

R 56V (Reichskulturkammer)
68, 95, 99.

R 57 (Deutsches Ausland-Institut)
169, 170, 221, 439, 674, 751.

R 57neu (Deutsches Ausland-Institut)
439, 1005, 1007.

R 58 (Reichssicherheitshauptamt)
148, 149, 158.

R 73 (Deutsche Forschungsgemeinschaft)
16405.

R 153 (Publikationsstelle-Berlin)
70, 71, 72, 80, 92, 94, 96, 110, 111, 115, 136, 233, 256, 257, 258, 277, 280, 282, 283, 289, 290, 349, 373, 374, 378, 379, 386, 436, 614, 623, 708, 710, 740, 745, 748, 768, 1068, 1070, 1114, 1143, 1190, 1274, 1299, 1305, 1345, 1424, 1478, 1529, 1559, 1629, 1709.

ZSg (Zeitgeschichtliche Sammlung)
1–142/19, 49.

Document Center, Berlin (BDC)

Personal files
Emil Meynen
Dora Nadge
Rupert von Schumacher
Karl Springenschmid
Arnold Hillen Ziegfeld

Forschungsprojekt, "Geographiegeschichte 1918–1939," Archiv, Universität Hamburg, Wirtschaftsgeographische Abteilung

RWTH Aachen, Fakultät I, Dekan to Minister für Wissenschaft, Kunst und Volksbildung, 8 November 1934, Anlage 1: Arbeitsbericht von Overbeck, 15 October 1934.

Sandner, G. and M. Roessler, *Endbericht für das Forschungsprojekt*, 1989.

Unpublished theses (*Staatsexamensarbeiten*) by Waldmann (1984), Gross (1976), and Böge (1987). See bibliography for full references.

Georg Westermann Werkarchiv (WWA)

Dehmel, Richard. "Die Geschichte der Kartographie im Haus Westermann durch 125 Jahre." Unpublished manuscript, no date.

Schulverlag IV.

Diercke Atlas Oberstufe, Neubearbeitung 1933.

v30/11, v30/217.

Reports on meetings of the *Deutscher Geographentag* etc., by Westermann representative H. Heyde.

Politisches Archiv des Auswärtigen Amtes, Bonn (AA)

Kult VI A, Deutschtum im Ausland-2 (DiA-2)
Nr. 11, Stiftung für deutsche Volks- und Kulturbodenforschung, Band 1–8.
Volksstiftung (Volz-Stiftung), Band 1–6.
VDA, Verein für das Deutschtum im Ausland, Band 1–17.
adh, Volksbund für das Deutschtum im Ausland, Band 1–5.
Stiftung Volk und Reich, Band 1.
Deutscher Schutzbund, Band 9.
Institut für Grenz- und Auslandstudien, Band 1.

Abteilung Inland
I, Partei, Propaganda, Band 1–6 (Paket 37, 38).
IIc, Ki.1, D 2-VFG (Volksdeutsche Forschungsgemeinschaften).
IIc, Ki.1, D 2-VDA (Volksbund für das Deutschtum im Ausland).
IIc, Ki.1, D 2-VFG (Historisch-geographischer Atlas von Europa).
IIc, Ki.1, D 3-Handwörterbuch für das Grenz- und Auslanddeutschtum.
IIg, 288, Volkstum und Grenzfragen.
Referat Deutschland, PO 25, Deutschtum im Auslande, Grenzlanddeutschtum, Kulturautonomie der Minderheiten, Band 1.
Referat Deutschland, PO 26, Politische und kulturelle Propaganda, Band 1, 2.

Other
Politische Abteilung, Pol. XVII, Geheim, Band 3.
Abt. II (Politik 2), Dt.-österreichischer Volksbund – Österreich.
R 23076.
R 27634.
R 30734.

Presseabteilung
Az: Hefte, Monatshefte *Volk und Reich*, Allgemeiner Schriftwechsel (Paket 482).
Deutschland 6, Volk und Reich, Band 1.
Deutschland 8, Presse, Propaganda, allgemein, Band 1–9.
Deutschland 19, Geheim, Reichszentrale für Heimatdienst (RfH), Band 1–7.
Handakten des Gesandschaftsrats Starke, Nr. 58–62 (Zeitschrift *Volk und Reich*).
P16, Verbreitung von Druckschriften.
P16, Bildmaterial.
P16f, Zusammenfassung der deutschen Werbe- und Aufklärungsarbeit.
P16c, Propaganda im Inland.

PRIMARY SOURCES

Public Record Office, Kew/London (PRO)

FO 1031/140   "Der Drang nach Osten," unpublished manuscript by E. Meynen, E. Otremba, W. Pillewizer, S. Schneider, A. Sievers, 21 April 1947.

B. PUBLISHED WORKS IN SPECIALIZED COLLECTIONS

1. *American Geographical Society Collection, Milwaukee (AGS)*
   The map and atlas collections of this important cartographic depository were examined, especially the editions of German atlases between ca. 1900 and 1945.

2. *Bundeszentrale für Politische Bildung, Archiv, Bonn (BfPB)*
   Publications of the Weimar *Reichszentrale für Heimatdienst*, such as, "Richtlinien," pamphlets, books, and the near complete collection of the journal *Der Heimatdienst*.

3. *Georg-Eckert-Institut für Internationale Schulbuchforschung, Braunschweig (GEI)*
   Collection of textbook editions for the subjects of geography and history.

   Editions of *Putzgers Historischer Weltatlas*, 42. Auflage 1920–52. Auflage 1935. Bielefeld, Leipzig: Verlag Velhagen & Klasing (incomplete collection).

4. *Georg Westermann Werkarchiv (WWA)*
   Editions of *Diercke Schulatlas, Grosse Ausgabe*, 51. Auflage 1918–82. Auflage 1943–44. Braunschweig: Georg Westermann Verlag (complete collection).

5. *Institut für Auslandsbeziehungen, Stuttgart* (formerly *Deutsches Ausland-Institut*).
   The library of this institute comprises one of the most extensive collections on German national issues in the world.

C. PERIODICAL SURVEY

The following periodicals were checked for maps to acquire an overview of trends:

Der Heimatdienst (1920–32)

Illustrierter Beobachter (1929–40)

Nationalsozialistische Monatshefte (1930–44)

Richtlinien der Reichszentrale für Heimatdienst (1926–32)

Schriften zur Geopolitik (vol. 1, 1932–vol. 22, 1943)

Der Schulungsbrief (1934–43)

Volk und Reich (1933–44)

Zeitschrift für Geopolitik (vol. 1, 1924–vol. 21, 1944)

Zeitschrift für Geopolitik, Beihefte (vol. 1, 1924–vol. 14, 1942)

D. INTERVIEWS AND CORRESPONDENCE

Few of the people involved in the mapping of German national territory were still alive or willing to participate when I carried out the original research. I was only able to obtain a minuscule amount of written and oral testimony and when I compared the information with other source material, it revealed the limitations of this "oral history." The persons either lacked critical distance, attempted to excuse their own involvement, or were incapable of making precise statements. I initiated correspondence with Wolfgang Höpker of Bonn, Germany, a member of the editorial board of the atlas *The War in Maps* (Wirsing 1941), and interviewed Richard Hartshorne in Madison, Wisconsin. Hartshorne had spent some time in the late 1920s at the *Stiftung für deutsche Volks- und Kulturbodenforschung*, an institute engaged in the coordination of the mapping of German national territory. I was also able to conduct an indirect interview with Dr Woldan of Vienna, Austria, who was involved in a national atlas project in the Nazi period. Dr Woldan refused to talk to strangers, but Dr M. Lazar of Vienna kindly acted as a mediator and asked specific questions, which I had sent to her in advance.

BIBLIOGRAPHY

Abel, Herbert (1935). Neuere Atlanten zur deutschen Landeskunde. *Geographischer Anzeiger* 36: 558–42.

Albert, Wilhelm (1935). *Grenz- und Auslanddeutsche Unterrichtsskizzen. Theoretische Einführung. Praktische Durchführung an Beispielen.* Leipzig: E. Wunderlich.

America to Speak in her Own Voice at the Peace Table (1917). *New York Times*, 29 September, pp. 1–2.

The American Geographical Society's Contribution to the Peace Conference (1919). *The Geographical Review* 7: 1–10.

Amtlicher Führer durch die Ausstellung Deutschland, Berlin 1936, 18. Juli bis 16. August (1936). Berlin: Gemeinnützige Berliner Ausstellungs-, Messe- und Fremdenverkehrs-G.m.b.H.

Anderson, Benedict (1991). *Imagined Communities. Reflections on the Origins and Spread of Nationalism.* London: Verso.

Anderson, James (1988). Nationalist Ideology and Territory. In *National Self-Determination and Political Geography*, ed. R. J. Johnston, David B. Knight, and Eleonore Kofman, 18–39. London: Croom Helm.

Andrees Allgemeiner Handatlas (1914, 1921, 1930). Bielefeld, Leipzig: Verlag Velhagen & Klasing.

Baldwin, Peter, ed. (1990). *Reworking the Past. Hitler, the Holocaust, and the Historians' Debate.* Boston: Beacon Press.

Banse, Ewald (1932). *Raum und Volk im Weltkriege. Gedanken über eine nationale Wehrlehre.* Oldenburg: Gerhard Stalling.

Bassin, Mark (1987a). Imperialism and the Nation State in Friedrich Ratzel's Political Geography. *Progress in Human Geography* 11,4: 473–95.

——— (1987b). Race Contra Space: The Conflict between German *Geopolitik* and National Socialism. *Political Geography Quarterly* 6: 115–34.

Batowski, Henryk (1970). Schwierige Nachbarschaft. In *Europa und die Einheit Deutschlands. Eine Bilanz nach Hundert Jahren*, ed. Walther Hofer, 181–200. Cologne: Wissenschaft und Politik.

Beck, Hanno (1982). *Große Geographen: Pioniere, Aussenseiter, Gelehrte.* Berlin: Reimer.

Belic, A. (1918). *Les cartes éthnographiques et historiques au service de la propaganda bulgare.* Questions Contemporaines 5. Paris: Ligue des Universitaires de Serbie.

Bessel, R. (1980). The Rise of the NSDAP and the Myth of Nazi Propaganda. *Wiener Library Bulletin*, vol. 33, no. 51/2: 20–9.

Blakemore, M. J. and J. B. Harley (1980). Concepts in the History of Cartography. A Review and Perspective. *Cartographica* 17, 4 (monograph 26).

Blumenwitz, Dieter (1980). *Die Darstellung der Grenzen Deutschlands in kartographischen Werken. Zur Verpflichtung zum Gebrauch verfassungskonformer Bezeichnungen durch die*

223

deutschen Behörden, insbesondere in Bezug auf die Belange der Ostdeutschen. Bonn: Kulturstiftung der deutschen Vertriebenen.

——— (1989). *Was ist Deutschland? Staats- und völkerrechtliche Grundsätze zur deutschen Frage und ihre Konsequenzen für die deutsche Ostpolitik.* Bonn: Kulturstiftung der deutschen Vertriebenen.

Boehm, Max Hildebert (1930). *Die deutschen Grenzlande.* Berlin: Reimar Hobbing.

——— (1932). *Das eigenständige Volk. Volkstheoretische Grundlagen der Ethnopolitik und Geisteswissenschaften.* Göttingen: Vandenhoeck & Ruprecht.

Boehm, Max Hildebert and A. H. Ziegfeld (1925). Mitteleuropas umstrittene Gebiete. In *Volk unter Völkern*, Bücher des Deutschtums 1, ed. Karl C. von Loesch and A. H. Ziegfeld, 242–4. Breslau: F. Hirt.

Böge, Wibeke (1987). Mitteleuropa: Konzepte, Abgrenzungen und inhaltliche Definition in Bezug auf die gegenwärtige Diskussionen. Staatsexamensarbeit, University of Hamburg.

Bohemian National Alliance of America (1917). Dismemberment of Austria. *The Bohemian Review* 1,1: 7–10.

Bowman, Isaiah (1922). *The New World. Problems in Political Geography.* Yonkers-on-Hudson, New York: World Book Co.

——— (1928). *The New World. Problems in Political Geography*, 4th ed. Yonkers-on-Hudson, New York: World Book Co.

Bracher, K. D. (1979). *Die deutsche Diktatur. Entstehung, Struktur, Folgen des Nationalsozialismus.* Frankfurt: Ullstein.

Bracher, K. D., M. Funke, and H. A. Jacobsen, eds. (1986). *Nationalsozialistische Diktatur 1933–1945. Eine Bilanz.* Bonn: Bundeszentrale für politische Bildung, Band 192.

——— eds. (1988). *Die Weimarer Republik 1918–1933. Politik, Wirtschaft, Gesellschaft.* Bonn: Bundeszentrale für politische Bildung, Band 251.

Braun, F. and A. H. Ziegfeld (1927). *Geopolitischer Geschichtsatlas.* Dresden: L. Ehlermann.

Braunias, Karl (1925). Zwei Jahre Minderheiteninstitut an der Wiener Universität. In *Volk unter Völkern*, Bücher des Deutschtums 1, ed. Karl C. von Loesch and A. H. Ziegfeld, 407–10. Breslau: F. Hirt.

Breuilly, John (1992). The National Idea in Modern German History. In *The State of Germany. The National Idea in the Making, Unmaking, and Remaking of a Modern Nation-State*, ed. J. Breuilly, 1–28. New York: Longman.

——— (1993). *Nationalism and the State*, 2nd ed. Chicago: University of Chicago Press.

Budding, Karl (1932). *Der polnische Korridor als europäisches Problem.* Danzig: Danziger Verlagsgesellschaft.

Burleigh, Michael (1988). *Germany Turns Eastwards: A Study of "Ostforschung" in the Third Reich.* Cambridge: Cambridge University Press.

Ciolina, Evamaria (1986). *Reklamesammelbilder. Massenmedien der NS-Zeit. Materialien zur volkstumsideologischen Durchdringung des Alltags, 1933–1945.* Munich: Institut für deutsche und vergleichende Volkskunde an der Ludwig-Maximilians-Universität (catalog accompanying exhibition of the conference on "*Volkskunde und Nationalsozialismus,*" 23–25 October 1986).

Clemenceau, Georges (1930). *Grandeur and Misery of Victory.* Trans. F. M. Atkinson. New York: Harcourt, Brace & Co.

Cohen, Saul B. (1991). Global Geopolitical Change in the Post-Cold War Era. *Annals of the Association of American Geographers* 81: 551–80.

Columbus Weltatlas. E. Debes Grosser Handatlas, Jubiläumsausgabe (1937). Berlin, Leipzig: Columbus Verlag, Paul Oestergaard K.G.

Commission Polonaise des Travaux Préparatoires au Congrès de la Paix (1919). *Question relatives aux territoires polonais.* Paris: privately printed.

Connor, Walker (1992). The Nation and its Myth. In *Ethnicity and Nationalism*, ed. A. D. Smith, 48–57. International Studies in Sociology and Social Anthropology 60. New York: E. J. Brill.

Czajka, W. et al. (1933). *Die deutsche Ostmark im Unterricht.* Breslau: Heinrich Handels Verlag.

Dahbour, Omar and Micheline R. Ishay, eds (1995). *The Nationalism Reader.* Atlantic Highlands, NJ: Humanities Press International.

Day, Clive (1921). The Atmosphere and Organization of the Peace Conference. In *What Really Happened at Paris. The Story of the Peace Conference, 1918–19*, ed. E. M. House and C. Seymour, 15–36. New York: Charles Scribner's Sons.

Demandt, Alexander (1990). Die Grenzen in der Geschichte Deutschlands. In *Deutschlands Grenzen in der Geschichte*, ed. Alexander Demandt, 9–31. Munich: C. H. Beck.

Der deutsche Volks- und Kulturboden (1925). *Deutsche Arbeit* 25,2: 87–8.

Diercke Schulatlas für höhere Lehranstalten. Grosse Ausgabe (1918–43/4). Braunschweig: Georg Westermann Verlag.

Dietrich, Bruno (1921). *Die natürliche Grenze Oberschlesiens.* Breslau: Grass, Barth & Co. W. Friedrich.

Diner, Dan (1984). "Grundbuch des Planeten": Zur Geopolitik Karl Haushofers. *Vierteljahrshefte für Zeitgeschichte* 32: 1–28.

Dmowski, Roman (1909). *La Question polonaise.* Paris: Armand Collin.

Dominian, Leon (1917). *The Frontiers of Language and Nationality in Europe*, American Geographical Society Special Publications 3. New York: Henry Holt.

Doubek, Franz (1938). Die Nationalitätenverhältnisse im Olsa-Schlesien. *Jomsburg* 2: 489–99.

Doubek, Franz and Erna Horn (1938). Die zahlenmässige Verbreitung des Deutschtums in Mittelpolen dargestellt in Gemeindeeinheiten, 1:750,000. *Jomsburg* 2: map appendix.

Eckert, Max (1921). *Die Kartenwissenschaft, Erster Band.* Berlin, Leipzig: Walter de Gruyter.

——— (1925). *Die Kartenwissenschaft, Zweiter Band.* Berlin, Leipzig: Walter de Gruyter.

Edney, Matthew (1993). Cartography Without "Progress": Reinterpreting the Nature and Historical Development of Mapmaking. *Cartographica* 30, 2/3: 54–68.

Eine unmögliche Genfer "Mitteleuropa"-Karte (1938). *Auslandsdeutsche Volksforschung*, 87–9.

Eley, Geoff (1980). *Reshaping the German Right. Radical Nationalism and Political Change after Bismarck.* New Haven, CT: Yale University Press.

Ernst, R. (1927). *Elsass. Taschenbuch des Grenz- und Auslanddeutschtums*, ed. K. C. von Loesch et al., vol. 7. Berlin: Deutscher Schutzbundverlag.

Eyck, Erich (1954). *Geschichte der Weimarer Republik*, 2 vols. Erlenbach/Zurich, Stuttgart: Eugen Rentsch.

Faber, Karl-Georg (1982). Zur Vorgeschichte der Geopolitik. Staat, Nation und Lebensraum im Denken deutscher Geographen vor 1914. In *Weltpolitik, Europagedanke, Regionalismus*, Festschrift für Heinz Gollwitzer zum 60. Geburtstag, ed. H. Dollinger et al., 389–406. Münster: Aschendorff.

Fahlbusch, Michael (1994). *"Wo der deutsche . . . ist, ist Deutschland." Die Stiftung für deutsche Volks- und Kulturbodenforschung in Leipzig 1920–1933.* Abhandlungen zur Geschichte der Geowissenschaften und Religion-Umwelt-Forschung, Beiheft 6. Bochum: Universitätsverlag Dr N. Brockmayer.

Fahlbusch, Michael, Mechthild Roessler, and Dominik Siegrist (1989). Conservatism, Ideology and Geography in Germany 1920–1950. *Political Geography Quarterly* 8: 353–67.

Fischer, Fritz (1977). *Griff nach der Weltmacht. Die Kriegszielpolitik des kaiserlichen Deutschland 1914/18*. Kronberg: Athenäum Verlag.

———— (1986). *From Kaiserreich to Third Reich. Elements of Continuity in German History, 1871–1945*. Trans. Roger Fletcher. London: Allen & Unwin.

Fittbogen, Gottfried (1924). *Was jeder Deutsche vom Grenz- und Auslanddeutschtum wissen muß*. Munich, Berlin: R. Oldenbourg.

———— (1927). *Wie lerne ich die Grenz- und Auslanddeutschen kennen? Einführung in die Literatur über die Grenz- und Auslanddeutschen*. Munich, Berlin: R. Oldenbourg.

———— (1928). *Die Deutschen außerhalb der Reichsgrenzen*. Reichszentrale für Heimatdienst, Richtlinie 70, September 1928. Berlin: Reichszentrale für Heimatdienst.

———— (1930). *Franz Xaver Mitterer und die Anfänge der Volkstumsarbeit*. Munich: C. H. Beck.

Folkers, Johann Ulrich (1937). *24 Karten zur Rassen- und Raumgeschichte des deutschen Volkes*. Berlin: J. Beltz.

Franke, Kurt F. K. (1989). Medien im Geschichtsunterricht der nationalsozialistischen Schule. In *Schule und Unterricht im Dritten Reich*, ed. Reinhard Dithmar, 59–85. Neuwied: Luchterhand.

Freilich, Josef (1918). *La Structure nationale de la Pologne*. Neuchâtel: privately printed.

Frenzel, K. (1938). Stellung und Aufgaben der kartographischen Privatindustrie. *Mitteilungen des Reichsamts für Landesaufnahme* 14: 365–76.

Frobenius, Leo (1921). *Atlas Africanus*, vol. 1. Munich: C. H. Beck.

Gandenberger von Moisy, F. (1933). Luftempfindlichkeit und geopolitische Forderungen. *Zeitschrift für Geopolitik* 10: 637–46.

Gea Verlag (1919). *Die Zerstückelung Deutschlands*. Berlin: Gea Verlag.

Geisler, Walter (1926). Politik und Sprachenkarten. *Zeitschrift für Geopolitik* 3: 701–13.

———— (1932a). *Schlesien als Raumorganismus*. Zur Wirtschaftsgeographie des deutschen Ostens, vol. 1, ed. Walter Geisler. Breslau: M. & H. Marcus.

———— ed. (1932b). *Wirtschafts- und verkehrsgeographischer Atlas von Oberschlesien*. Breslau: M. & H. Marcus.

———— (1933). Die Sprachen- und Nationalitätenverhältnisse an den deutschen Ostgrenzen und ihre Darstellung. *Petermanns Geographische Mitteilungen, Ergänzungshefte*, vol. 21. Gotha: Justus Perthes.

———— (1934). Die Problematik der Völker- und Sprachenkarten dargelegt am Beispiel der deutschen Ostgrenze. *Petermanns Geographische Mitteilungen* 80: 339–40

———— (1939). Die polnische Lügenpropaganda und ihre Auswirkungen während der Diktatsverhandlungen in Versailles und St. Germain. *Volk und Reich* 15: 550–5.

Gelfand, Lawrence E. (1976). *The Inquiry. American Preparations for Peace, 1917–1919*. New Haven, CT: Yale University Press, 1963; reprint, Westport, CT: Greenwood Press (page references are to reprint edition).

Geographie und Nationalsozialismus. 3 Fallstudien zur Institution Geographie im Deutschen Reich und der Schweiz (1989). Urbs et Regio 51, Kasseler Schriften zur Geographie und Planung, Kassel: Gesamthochschule Kassel.

Gerson, Louis L. (1953). *Woodrow Wilson and the Rebirth of Poland, 1914–1920*. New Haven, CT: Yale University Press.

Gottmann, Jean (1973). *The Significance of Territory*. Charlottesville: University Press of Virginia.

Gradmann, Robert (1929). *Wörterbuch deutscher Ortsnamen in den Grenz- und Auslands-gebieten*. Stuttgart: Verlag Ausland und Heimat.

Grimm, Bruno (1939). *Gau Schweiz. Dokumente über die nationalsozialistischen Umtriebe in der Schweiz*. Zurich: Jean Christophe Verlag (Sozialdemokratische Partei, Schweiz: Kultur und Arbeit, Schriften zur Wirtschafts-, Sozial- und Kultur-politik).

Gross, Wolfgang (1976). Nationalsozialistische Tendenzen im Erdkundeunterricht von 1933–1945 unter besonderer Berücksichtigung der Lehrbücher: Die Dar-stellung Deutschlands. Staatsexamensarbeit, University of Trier.

Grupp, Peter (1988). Vom Waffenstillstand zum Versailler Vertrag. In *Die Weimarer Republik 1918–1933. Politik, Wirtschaft, Gesellschaft*, ed. K. D. Bracher, M. Funke, and H. A. Jacobsen, 285–302. Bonn: Bundeszentrale für politische Bildung, Band 251.

Günther, Hans F. K. (1930). *Rassenkunde des deutschen Volkes*, 14th ed. Munich: J. F. Lehmanns Verlag

Gürtler, Arno (1931). *Das Zeichnen im erdkundlichen Unterricht*. Leipzig: E. Wunder-lich.

Häberle, D. (1919). Der Anteil der Deutschen und Polen an der Bevölkerung von West Preussen und Posen (nach A. Penck). *Geographische Zeitschrift* 25: 124–7.

Hagemeyer, Hans and Georg Leibbrandt, eds. (1939). *Europa und der Osten*. Munich: Hoheneichen Verlag.

Hall, Derek R. (1981). A Geographical Approach to Propaganda. In *Political Studies from Spatial Perspectives*, ed. Alan D. Burnett and Peter J. Taylor, 313–30. New York: John Wiley & Sons.

Hamerow, Theodore S. (1983). Review Essay. Guilt, Redemption, and Writing German History. *The American Historical Review* 88: 53–72.

Harley, J. B. (1988a). Maps, Knowledge, and Power. In *The Iconography of Landscape*, ed. Denis Cosgrove and Stephen Daniels, 277–312. Cambridge: Cambridge University Press.

——— (1988b). Silences and Secrecy: The Hidden Agenda of Cartography in Early Modern Europe. *Imago Mundi* 40: 57–76.

——— (1989). Deconstructing the Map. *Cartographica* 26,2: 1–20.

——— (1990). Cartography, Ethics and Social Theory. *Cartographica* 27,2: 1–23.

——— (1992). Deconstructing the Map. In *Writing Worlds. Discourse, Text and Metaphor in the Representation of Landscape*, ed. Trevor J. Barnes and James S. Duncan, 231–47. London: Routledge.

Harley, J. B. and David Woodward (1987). Concluding Remarks. In *The History of Cartography*. vol. 1, *Cartography in Prehistoric, Ancient, and Medieval Europe and the Mediterranean*, ed. J. B. Harley and David Woodward, 502–9. Chicago: University of Chicago Press.

Hartshorne, Richard (1935). Recent Developments in Political Geography. *Amer-ican Political Science Review* 29: 785–804, 943–66.

Hassinger, Hugo (1919). Bemerkungen über die Südostgrenze des deutschen Sprachgebiets. *Geographische Zeitschrift* 25: 215–19.

Haushofer, Karl (1922). Die suggestive Karte. *Grenzboten* 81 (reprinted as Haus-hofer 1928).

——— (1927). *Grenzen in ihrer geographischen und politischen Bedeutung*. Berlin: K. Vowinckel.

——— (1928). Die suggestive Karte. In *Bausteine zur Geopolitik*, ed. Karl Haus-hofer et al., 343–48. Berlin: K. Vowinckel.

——— (1932). Rückblick und Vorschau auf das geopolitische Kartenwesen. *Zeitschrift für Geopolitik* 9: 735–45.

———— (1933). Fromme Wünsche. Die slawische Idee der Absperrung des Deutschtums vom Osten. *Zeitschrift für Geopolitik* 10: 330–3.

———— (1939). *Grenzen in ihrer geographischen und politischen Bedeutung.* Heidelberg, Berlin: K. Vowinckel.

Haushofer, Karl and Kurt Trampler, eds (1931). *Deutschlands Weg an der Zeitwende.* Munich: Hugendubel.

Heinrich, Horst-Alfred (1990). Der politische Gehalt des fachlichen Diskurses in der Geographie Deutschlands zwischen 1920 und 1945 und dessen Affinität zum Faschismus. *Geographische Zeitschrift* 78: 209–26.

Heiss, Friedrich and A. H. Ziegfeld, eds (1933). *Deutschland und der Korridor.* Berlin: Volk und Reich.

Heiss, Friedrich, Rudolf Fischer, and Waldemar Wucher, eds (1938). *Die Wunde Europas. Das Schicksal der Tschecho-Slowakei.* Berlin: Volk und Reich.

Heiss, Friedrich, Günther Lohse, and Waldemar Wucher, eds (1939). *Deutschland und der Korridor.* Berlin: Volk und Reich.

Henrikson, A. K. (1975). The Map as an Idea: The Role of Cartographic Imagery during the Second World War. *The American Cartographer* 2,1: 19–53.

Herb, Guntram Henrik (1989). Persuasive Cartography in *Geopolitik* and National Socialism. *Political Geography Quarterly* 8: 289–303.

———— (1993). National Self-Determination, Maps, and Propaganda in Germany 1918–1945. Ph.D. diss., University of Wisconsin-Madison.

Herzstein, Robert Edwin (1978). *The War that Hitler Won. The Most Infamous Propaganda Campaign in History.* New York: G. P. Putnam's Sons.

Heske, Hennig (1986). Political Geographers of the Past III. German Geographical Research in the Nazi Period: A Content Analysis of the Major Geography Journals. *Political Geography Quarterly* 5: 267–81.

———— (1988). *. . . morgen die ganze Welt: Erdkundeunterricht im Nationalsozialismus.* Giessen: Focus Verlag.

Heyde, Herbert (1919). Die Nationalitäten in den deutschen Ostprovinzen. Eine Fälschung schlimmster Art. *Zeitschrift der Gesellschaft für Erdkunde zu Berlin*, 185–86.

Hilgemann, Werner (1984). *Atlas zur deutschen Zeitgeschichte 1918–1968.* Munich: Piper.

Himstedt, A. (1939). Wehrwille und Wehrkraft. *Der Schulungsbrief* 6: 114–23.

Hitler, Adolf (1933). *Mein Kampf.* Munich: Franz Eher Nachfolger.

Hobsbawm, E. J. (1990). *Nations and Nationalism Since 1780. Programme, Myth, Reality,* 2nd ed. Cambridge: Cambridge University Press.

Holborn, Hajo (1965). *The Political Collapse of Europe.* New York: Alfred A. Knopf.

———— (1969). *A History of Modern Germany, 1840–1945.* Princeton, NJ: Princeton University Press.

Holdar, Sven (1992). The Ideal State and the Power of Geography. The Life-Work of Rudolf Kjellén. *Political Geography* 11: 307–23.

Horn, Erna (1939). Deutsches Memelland. *Jomsburg* 3: 73–8.

Hudson, Brian (1977). The New Geography and the New Imperialism: 1870–1918. *Antipode* 9,2: 12–19.

Humbel, Kurt (1977). *Nationalsozialistische Propaganda in der Schweiz 1931–1939.* Res Publica Helvetica 6. Bern, Stuttgart: Verlag Paul Haupt.

Isbert, Otto Albrecht (1937a). Volksbodenkarten. *Nation und Staat. Deutsche Zeitschrift für das europäische Minoritätenproblem* 10: 490–501.

———— (1937b). Zur kartographischen Darstellung des Auslandsdeutschtums. *Auslandsdeutsche Volksforschung* 1: 98–105.

—————— (1938). Statistik und Kartographie im Dienste der Volksforschung. *Auslandsdeutsche Volksforschung* 2: 151–61.

Istituto geografico de Agostini (1917). *L'Europe ethnique et linguistique. Atlas descriptif.* Novara: Istituto geografico de Agostini.

Jackall, Robert, ed. (1995). *Propaganda. Main Trends of the Modern World.* New York: New York University Press.

Jacobsen, Hans-Adolf (1970). *Hans Steinacher. Bundesleiter des VDA 1933–1937.* Schriften des Bundesarchivs 19. Boppard am Rhein: Harold Boldt Verlag.

—————— (1979). *Karl Haushofer. Leben und Werk.* 2 vols. Schriften des Bundesarchivs 24. Boppard am Rhein: Harold Boldt Verlag.

Jäger, Fritz (1924). Die deutsch–polnische Grenze. *Zeitschrift der Gesellschaft für Erdkunde zu Berlin,* 257–80.

—————— (1928). Die deutsch–polnische Grenze. Erörterung über Probleme der Grenzziehung. *Friedrich Mann's pädagogisches Magazin,* no. 1194 (Deutscher Staat: Schriften zur politischen Bildung, V. Reihe, Grenzlande, Heft 9).

Jantzen, Walther (1936a). Der Geographentag gegen kindliche Bildkarten. *Zeitschrift für Geopolitik* 13: 841–3.

—————— (1936b). Kartenplakate für Aufklärung und Werbung. *Zeitschrift für Geopolitik* 13: 696–700.

—————— (1937). Das Gepräge des deutschen Volkes auf den geopolitischen Wandkarten. *Zeitschrift für Geopolitik* 14: 670–4.

—————— (1938). Der Pfeil im Kartenbild. *Weltanschauung und Schule,* 318–28.

—————— (1942). Geopolitik im Kartenbild. *Zeitschrift für Geopolitik* 19: 353–8.

—————— (1943). *Geopolitik im Kartenbild: Verrat an Europa.* Berlin, Heidelberg: K. Vowinckel.

Jarausch, Konrad H. (1988). Removing the Nazi Stain? The Quarrel of the German Historians. *German Studies Review* 11: 285–301.

Jaworski, Rudolf (1978). Der auslandsdeutsche Gedanke in der Weimarer Republik. *Annali dell'Istituto storico italo-germanico in Trento* 4: 369–86.

—————— (1980). Kartographische Ortsbezeichnungen und nationale Emotionen: Ein deutsch–tschechischer Streitfall aus den Jahren 1924/25. In *Stadtverfassung, Verfassungsstaat, Pressepolitik. Festschrift für Eberhard Nanjoks zum 65. Geburtstag,* ed. Franz Quarthal and Wilfried Setzler, 250–61. Sigmaringen: Jan Thorbecke Verlag.

—————— (1984). Deutsch–polnische Feindbilder 1919–1932. *Internationale Schulbuchforschung* 6: 140–56.

Jessop, T. E. (1942). *The Treaty of Versailles. Was It Just?* London: Thomas Nelson & Sons.

Jezowa, Kazimiera (1933). *Politische Propaganda in der deutschen Geographie.* Danzig: Towarzystwo Przyjaciol Nauki i Sztuki w Gdansku E.V.

Johnston, R. J., D. B. Knight, and E. Kofman (1988). *Nationalism, Self-Determination and Political Geography.* London: Croom Helm.

Jowett, G. S. and V. O'Donnell (1986). *Propaganda and Persuasion.* London: Sage Publications.

Kaplan, David H. (1994). Two Nations in Search of a State: Canada's Ambivalent Spatial Identities. *Annals of the Association of American Geographers* 84: 585–606.

Keller, Karl (1929). *Die fremdsprachige Bevölkerung in den Grenzgebieten des Deutschen Reiches.* Begleitschrift zum Kartenwerk: Sprachenatlas der deutschen Grenzgebiete, herausgegeben von der Reichszentrale für Heimatdienst. Berlin: Zentralverlag.

Kershaw, Ian (1983a). How Effective was Nazi Propaganda? In *Nazi Propaganda. The Power and the Limitations,* ed. D. Welch, 180–205. London: Croom Helm.

——— (1983b). Ideology, Propaganda, and the Rise of the Nazi Party. In *The Nazi Machtergreifung*, ed. P. D. Stachura, 162–81. London: George Allen & Unwin.

——— (1989). *The Nazi Dictatorship. Problems and Perspectives of Interpretation*, 2nd ed. London: Edward Arnold.

Keyser, Erich (1919). *Westpreußen und das deutsche Volk. Nebst einer Bevölkerungskarte.* Danzig: A. W. Kasemann.

Kirchhoff, Alfred (1898). Langhans' deutscher Kolonial-Atlas. *Petermanns Geographische Mitteilungen* 44: 13–14.

Kjellen, Rudolf (1914). *Die Großmächte der Gegenwart.* Leipzig, Berlin: Teubner.

——— (1917). *Der Staat als Lebensform.* Trans. Margarethe Langfeldt. Leipzig: S. Hirzel.

Knieper, F. (1931). *Das Zeichnen im erdkundlichen Unterricht.* Leipzig: E. Wunderlich.

Knight, David B. (1984). Geographical Perspectives on Self-Determination. In *Political Geography. Recent Advances and Future Directions*, ed. Peter Taylor and John House, 168–90. Totowa, NJ: Barnes & Nobles.

Koehl, Robert Lewis (1957). *RKFDV. German Resettlement and Population Policy, 1939–1945. A History of the Reich Commission for the Strengthening of Germandom.* Cambridge, MA: Harvard University Press.

——— (1983). *The Black Corps. The Structure and Power Struggles of the Nazi SS.* Madison: University of Wisconsin Press.

Köhler, Franz (1987). *Gothaer Wege in Geographie und Kartographie.* Gotha: VEB Hermann Haack.

Komarnicki, Titus (1957). *Rebirth of the Polish Republic. A Study in the Diplomatic History of Europe, 1914–20.* London: William Heinemann.

Korherr, Richard (1938). *Volk und Raum.* Würzburg: Universitätsdruckerei H. Stürtz.

Korinman, Michel (1990). *Quand l'Allemagne pensait le monde. Grandeur et décadence d'une géopolitique.* Paris: Librairie Arthème Fayard.

Kost, Klaus (1986). Begriffe und Macht. Die Funktion der Geopolitik als Ideologie. *Geographische Zeitschrift* 74: 14–30.

——— (1988). *Die Einflüsse der Geopolitik auf Forschung und Theorie der politischen Geographie von ihren Anfängen bis 1945. Ein Beitrag zur Wissenschaftsgeschichte der politischen Geographie und ihrer Terminologie unter besonderer Berücksichtigung von Militär- und Kolonialgeographie.* Bonner Geographische Abhandlungen, vol. 76. Bonn: Ferd. Dümmlers Verlag.

Krebs, Norbert (1937). *Atlas des deutschen Lebensraumes in Mitteleuropa.* Leipzig: Bibliographisches Institut.

Kredel, O. and F. Thierfelder (1931). *Deutsch–fremdsprachiges Ortsnamenverzeichnis.* Berlin: Deutsche Verlagsgesellschaft.

Krieg, Hans (1939). Landkarten als Mittel der politischen Propaganda. *Zeitschrift für Politik* 29: 663–9.

Kristof, Ladis K. D. (1960). The Origins and Evolution of Geopolitics. *Journal of Conflict Resolution* 4: 15–51.

Kronacher, Bettina (1938). *Das Reich der Deutschen. Deutsche Geschichte in Umrissen.* vol. 1, *Der deutsche Lebensraum in der Geschichte.* Frankfurt: Moritz Diesterweg.

Kuffner, Hanus (1922). *Unser Staat und der Weltfrieden.* Trans. H.V., Vienna. Warnsdorf in Böhmen: Ed. Strache.

Lacoste, Yves (1986). Géopolitiques de la France. *Hérodote* 40: 5–31.

Lange, Friedrich (1925). Das Rheinland und der deutsche Osten. *Der Heimatdienst* 5,4: 51–2.

——— (1937). *Volksdeutsche Kartenskizzen.* Berlin: Volksbund für das Deutschtum im Ausland.

———— (1940). *Wir zwischen 25 Nachbarvölkern*. Berlin: Verlag der Deutschen Arbeitsfront.

Langewiesche, Dieter (1992a). Germany and the National Question in 1848. In *The State of Germany. The National Idea in the Making, Unmaking, and Remaking of a Modern Nation-State*, ed. J. Breuilly, 60–79. New York: Longman.

———— (1992b). Reich, Nation und Staat in der jüngeren deutschen Geschichte. *Historische Zeitschrift* 254: 341–81.

Langhans, Paul (1905). *Alldeutscher Atlas*. Gotha: Justus Perthes.

———— (1916). Deutsche Landkarten unter dem Einfluss des Weltkriegs. *Petermanns Geographische Mitteilungen* 62: 379–80.

Langhans-Ratzeburg, Manfred (1929). Die geopolitischen Reibungsgürtel der Erde. *Zeitschrift für Geopolitik* 6: 158–67.

Lansing, Robert (1921). *The Peace Negotiations. A Personal Narrative*. Boston: Houghton Mifflin Co.

Laun, Rudolf (1919). *Die tschechoslowakischen Ansprüche auf deutsches Land*. Vienna: Alfred Hölder, Universitätsbuchhändler; reprint, Seeds of Conflict Series 6, Minorities in Czechoslovakia.

Lautensach, Hermann (1924). Geopolitik und staatsbürgerliche Bildung. *Zeitschrift für Geopolitik* 1: 467–76.

———— (1929). Ein geopolitischer Typenatlas. *Zeitschrift für Geopolitik* 6: 608–11.

Lebeck, Robert and Manfred Schütte (1980). *Propagandapostkarten I. 80 Bildpostkarten aus den Jahren 1898–1929*. Dortmund: Harenberg Kommunikation.

———— (1985). *Propagandapostkarten II. 80 Bildpostkarten aus den Jahren 1933–1943*. Dortmund: Harenberg Kommunikation.

LeBrun, Didier (1990). Allemagne et Pologne, ou la carte comme arme de propagande. *Mappemonde* 1: 14–16.

Leers, Johann von and Konrad Frenzel (1934). *Atlas zur deutschen Geschichte der Jahre 1914 bis 1933*. Bielefeld, Leipzig: Velhagen & Klasing.

Leggewie, Claus (1990). *Die Republikaner. Ein Phantom nimmt Gestalt an*. Berlin: Rotbuch Verlag.

Légrády, Otto (1930). *Justice for Hungary! The Cruel Errors of Trianon*. Budapest: Pesti Hirlap.

Lissmann, Hans Joachim (1989). Sachunterricht in der Grundschule. Eine Fallstudie für die erste Phase des Dritten Reiches. In *Schule und Unterricht im Dritten Reich*, ed. Reinhard Dithmar, 235–57. Neuwied: Luchterhand.

Loesch, Karl C. von (1925). Der Deutsche Schutzbund. Die Ziele. In *Volk unter Völkern*, Bücher des Deutschtums 1, ed. Karl C. von Loesch and Arnold Hillen Ziegfeld, 9–21. Breslau: F. Hirt.

———— (1937). Der burgenländische Korridor. Ein Misserfolg des tschechischen Imperialismus. *Volk und Reich* 13: 410–17.

Loesch, Karl C. von and Max Hildebert Boehm, eds. (1930). *Grenzdeutschland seit Versailles. Die grenz- und volkspolitischen Folgen des Friedensschlusses*. Berlin: Brückenverlag.

Loesch, Karl C. von and Arnold Hillen Ziegfeld, eds (1925). *Volk unter Völkern*. Bücher des Deutschtums 1. Breslau: F. Hirt.

———— eds (1926). *Staat und Volkstum*, Bücher des Deutschtums 2. Berlin: Deutscher Schutzbund Verlag.

Luckau, Alma (1941). *The German Delegation at the Paris Peace Conference*. New York: Columbia University Press.

Ludendorff, Erich (1920). *Meine Kriegserinnerungen 1914–1918*. Berlin: Ernst Siegfried Mittler & Sohn.

Lukas, Georg A. (1919). Geographie und völkische Schutzarbeit. *Geographische Zeitschrift* 25: 235–45.

Lundgreen, Peter, ed. (1985). *Wissenschaft im Dritten Reich*. Edition Suhrkamp 1306, Neue Folge 306. Frankfurt: Suhrkamp Verlag.

Lundgreen-Nielsen, Kay (1979). *The Polish Problem at the Paris Peace Conference. A Study of the Policies of the Great Powers and the Poles, 1918–1919*. Trans. Alison Borch-Johansen. Odense Studies in History and Social Sciences 59. Odense: Odense University Press.

Machatschek, Fritz (1928). *Die Tschechoslowakei*. Weltpolitische Bücherei, vol. 8. Berlin: Zentralverlag.

Mackinder, Halford, T. (1904). The Geographical Pivot of History. *Geographical Journal* 23,4: 421–44.

——— (1919). *Democratic Ideals and Reality. A Study in the Politics of Reconstruction*. London: Constable; New York: Holt.

Mantoux, Etienne (1952). *The Carthaginian Peace or the Economic Consequences of Mr. Keynes*. New York: Charles Scribner's Sons.

Martin, Geoffrey J. (1980). *The Life and Thought of Isaiah Bowman*. Hamden, CT: Archon Books.

März, Josef (1921). Die Landkarte als politisches Propagandamittel. *Die Gartenlaube* 16: 261–2.

Masaryk, Thomas G. (1917). The Future Status of Bohemia. *The Bohemian Review* 1,3: 1–8.

——— (1927). *The Making of a State. Memoirs and Observations, 1914–18*. New York: Frederick A. Stokes Co.

Materialien zur ostdeutschen Frage (1919). *Mitteilungen der Deutschen Volksräte Posens und Westpreussens*. 1,2: 2–5.

Mauersberger, Volker (1971). *Rudolf Pechel und die "Deutsche Rundschau". Eine Studie zur konservativ-revolutionären Publizistik in der Weimarer Republik (1918–1933)*. Studien zur Publizistik, Bremer Reihe, Deutsche Presseforschung 16. Bremen: Schünemann Universitätsverlag.

Maull, Otto (1925). *Politische Geographie*. Berlin: Gebrüder Borntraeger.

——— (1928). Über politischgeographische-geopolitische Karten. In *Bausteine zur Geopolitik*, ed. Karl Haushofer et al., 325–42. Berlin: K. Vowinckel.

Mayer, Tilman (1987). *Prinzip Nation. Dimensionen der nationalen Frage, dargestellt am Beispiel Deutschlands*. Forschungstexte, Wirtschafts- und Sozialwissenschaften 16. Opladen: Leske & Budrich.

Meissner-Hohenmeiss, F. (1937). Die Volkspolitik der Tschechen im Kartenbild. *Zeitschrift für Geopolitik* 14: 434–46.

Mehmel, Astrid (1990). Geographie und Versailler Vertrag. Reaktionen deutscher Geographen unter besonderer Berücksichtigung ihrer Stellungnahmen zu Fragen der "Grenzproblematik" und territorialen Veränderungen in Europa 1918–1933. Diplomarbeit, Institut für Wirtschaftsgeographie, University of Bonn.

——— (1995). Deutsche Revisionspolitik in der Geographie nach dem ersten Weltkrieg. *Geographische Rundschau* 47: 498–505.

Metz, Friedrich (1934). Eine Verkehrskarte von Deutschland. *Deutsche Arbeit* 34,7: 377–8.

Meyer, Henry Cord (1946). Mitteleuropa in German Political Geography. *Annals of the Association of American Geographers*, no. 9, 178–94.

——— (1955). *Mitteleuropa in German Thought and Action*. The Hague: M. Nijhoff.

Meynen, Emil (1935). *Deutschland und Deutsches Reich. Sprachgebrauch und Begriffswesenheit des Wortes Deutschland*. Leipzig: Brockhaus.

Mezes, Sidney Edward (1921). Preparations for Peace. In *What Really Happened at*

Paris. The Story of the Peace Conference, 1918–19, ed. Edward Mandell House and Charles Seymour, 1–14. New York: Charles Scribner's Sons.

Milleker, R. (1937). Über ethnographische Karten als Grundlage geopolitischer Entscheidungen. *Zeitschrift für Geopolitik* 14: 639–45.

Monmonier, Mark (1991). *How to Lie with Maps.* Chicago: University of Chicago Press.

Morris, W. B. (1982). *The Weimar Republic and Nazi Germany.* Chicago: Nelson-Hall.

Mosse, George L. (1975). *The Nationalization of the Masses. Political Symbolism and Mass Movements in Germany from the Napoleonic Wars Through the Third Reich.* New York: Howard Fertig.

——— (1981). *The Crisis of German Ideology. The Intellectual Origins of the Third Reich.* New York: Grosset & Dunlap, 1964; reprint, New York: Schocken Books (page references are to reprint edition).

Murphy, Alexander B. (1990). Historical Justifications for Territorial Claims. *Annals of the Association of American Geographers* 80: 531–48.

Murphy, David T. (1992). The Heroic Earth: The Flowering of Geopolitical Thought in Weimar Germany, 1924–1933. Ph.D. diss., University of Illinois at Urbana-Champaign.

Natter, Wolfgang, Theodore R. Schatzki, and John Paul Jones III, eds (1995). *Objectivity and its Other.* New York: Guilford Press.

Neumann, Franz (1944). *Behemoth. The Structure and Practice of National Socialism, 1933–1944.* New York: Oxford University Press.

Niedermayer, Oskar Ritter von (1939). *Wehrgeographischer Atlas von Frankreich.* Berlin: Reichsdruckerei.

——— (1941). *Wehrgeographischer Atlas der Union der Sozialistischen Sowjetrepubliken.* Berlin: Reichsdruckerei.

Norton, Donald Hawley (1965). Karl Haushofer and his Influence on Nazi Ideology and German Foreign Policy, 1919–1945. Ph.D. diss., Clark University.

Oberhummer, Eugen (1920). Die politische Karte Europas nach serbischen Plänen aus dem Anfang des Weltkrieges. *Petermanns Geographische Mitteilungen* 66: 190.

Oberkrome, Willi (1993). *Volksgeschichte. Methodische Innovationen und völkische Ideologisierung in der deutschen Geschichtswissenschaft 1918–1945.* Kritische Studien zur Geschichtswissenschaft 101. Göttingen: Vandenhoeck & Ruprecht.

Obst, Erich (1929). Oberschlesien. *Zeitschrift für Geopolitik* 6: 756–71.

Oerdingen, H. von and A. Stein (1934). *Deutsches Volk, Grenzen zerschneiden Dich!* Cologne: Bachem.

Overbeck, Hermann and Georg Sante (1934). *Saaratlas.* Gotha: Justus Perthes.

Parker, Geoffrey (1985). *Western Geopolitical Thought in the Twentieth Century.* New York: St. Martin's Press, 1985.

Paul, Gerhard (1990). *Aufstand der Bilder. Die NS-Propaganda vor 1933.* Bonn: Verlag J. H. W. Dietz Nachf.

Penck, Albrecht (1919a). Deutsche und Polen in West-Preussen und Posen. *Deutsche Allgemeine Zeitung*, no. 67, 9 February 1919, Morgenausgabe, Beiblatt.

——— (1919b). Die Polengrenze. *Illustrierte Zeitung, Leipzig*, vol. 152, no. 3960, 22 May 1919, pp. 536–7.

——— (1919c). Verteilung der Deutschen und Polen in Westpreussen und Posen. *Zeitschrift der Gesellschaft für Erdkunde zu Berlin*, Karte 1.

——— (1921a). Die Deutschen im Polnischen Korridor. *Zeitschrift der Gesellschaft für Erdkunde zu Berlin*, 169–85.

——— (1921b). Oberschlesien deutsch oder polnisch? *Deutsche Allgemeine Zeitung*, no. 216, 11 May 1921, Morgenausgabe, Beiblatt.

——— (1925). Deutscher Volks- und Kulturboden. In *Volk unter Völkern*, Bücher

233

des Deutschtums 1, ed. Karl C. von Loesch and A. H. Ziegfeld, 62–73. Breslau: F. Hirt

Penck, Albrecht and Hans Fischer (ca. 1925). *Der deutsche Volks- und Kulturboden in Europa*, 1: 3,270,000. Berlin: Verein für das Deutschtum im Ausland.

Penck, Albrecht and Herbert Heyde (1919). *Karte der Verbreitung von Deutschen und Polen längs der Warthe–Netze-Linie und der unteren Weichsel*, 1:100,000. Berlin: Gea Verlag.

———— (1921). Die Deutschen im "Polnischen Korridor." Karte der Verbreitung der Deutsch- und Polnisch-Sprechenden auf Grund der Volkszählung vom 1. Dezember 1910, 1:300,000. (Berlin: Preussische Landesaufnahme). *Zeitschrift der Gesellschaft für Erdkunde zu Berlin*: appendix.

Pessler, Wilhelm (1931). *Deutsche Volkstumsgeographie*. Braunschweig: Verlag Georg Westermann.

Pickles, John (1992). Texts, Hermeneutics and Propaganda Maps. In *Writing Worlds. Discourse, Text and Metaphor in the Representation of Landscape*, ed. Trevor J. Barnes and James S. Duncan, 193–230. London: Routledge.

Pomoranus (ca. 1931). *Das Deutschtum in Westpreussen und Posen*. Taschenbuch des Grenz- und Auslanddeutschtums, ed. Karl C. von Loesch et al., vol. 17/18. Berlin: Deutscher Schutzbundverlag.

Possekel, Kurt (1986). Verein für das Deutschtum im Ausland (VDA). In *Lexikon zur Parteiengeschichte. Die bürgerlichen und kleinbürgerlichen Parteien und Verbände in Deutschland (1789–1945)*, ed. Dieter Fricke et al., vol. 4, 282–97. Cologne: Pahl-Rugenstein; Leipzig: Bibliographisches Institut.

Praesent, H. (1919a). Deutsches und polnisches Sprachgebiet. *Geographische Zeitschrift* 25: 219–22.

———— (1919b). Review of *Sprachenkarte der deutschen Ostmarken* by Dietrich Schäfer and *Nationalitätenkarte der östlichen Provinzen des deutschen Reiches nach den Ergebnissen der amtlichen Volkszählung vom Jahre 1910* by Jakob Spett. In *Geographische Zeitschrift*, 128–9.

Prestwick, Roger (1978). Maps and the Perception of Space. In *An Invitation to Geography*, ed. D. Lanegran and R. Palms, 13–37. New York: McGraw-Hill.

Pudelko, Alfred and A. H. Ziegfeld (1938). *Kleiner deutscher Geschichtsatlas*. Berlin: Edwin Runge.

F. W. Putzgers Historischer Schulatlas. Grosse Ausgabe (1920–35). Leipzig, Bielefeld: Verlag Velhagen & Klasing.

Quam, L. O. (1943). The Use of Maps in Propaganda. *Journal of Geography* 42: 21–32.

Rado, Alex (1980). *Atlas für Politik, Wirtschaft, Arbeiterbewegung. I. Imperialismus*. Vienna, Berlin: Verlag für Literatur und Politik 1930; reprint, Gotha: VEB Hermann Haack (page references are to reprint edition).

Ratzel, Friedrich (1897). *Politische Geographie: oder, die Geographie der Staaten, des Verkehrs und des Krieges*. Munich: Oldenbourg.

Reichszentrale für Heimatdienst (1919). *Deutschlands Versklavung. Das Dokument des Unfriedens*. Berlin: Zentralverlag.

———— (1929). *Sprachenatlas der Grenzgebiete des Deutschen Reiches nach den Ergebnissen der Volkszählung vom 16. IV. 1925*. Berlin: Zentralverlag.

Reinhard, R. (1920). *Die Welt nach dem Friedensschluss. Ein geographisch-wirtschaftspolitischer Überblick*. Breslau: F. Hirt.

Rhodes, Anthony (1976). *Propaganda. The Art of Persuasion: World War II*. New York: Chelsea House.

Ribaric, J., F. de Sisic, and N. Zic (1919). *La Question de l'Adriatique. L'Istrie*. Paris.

Rich, Norman (1973). *Hitlers War Aims.* Vol. 1, *Ideology, The Nazi State, and the Course of Expansion.* New York: W. W. Norton & Co.

——— (1974). *Hitlers War Aims.* Vol. 2, *The Establishment of the New Order.* New York: W. W. Norton & Co.

Rieth, Adolf (1927). *Die geographische Verbreitung des Deutschtums in Rumpf-Ungarn in Vergangenheit und Gegenwart.* Schriften des Deutschen Ausland-Instituts Stuttgart, A. Kulturhistorische Reihe 18. Stuttgart: Ausland und Heimat Verlags-Aktiengesellschaft.

Ritter, Ernst (1976). *Das Deutsche Ausland-Institut in Stuttgart 1917–1945. Ein Beispiel der Volkstumsarbeit zwischen den Weltkriegen.* Frankfurter Historische Abhandlungen 14. Wiesbaden: Franz Steiner Verlag.

Rizoff, D., ed. (1917). *Die Bulgaren in ihren historischen, ethnographischen und politischen Grenzen. Atlas mit 40 Landkarten.* Berlin: Königliche Hoflithographie, Hof-, Buch- und Steindruckerei Wilhelm Greve.

Roessler, Mechthild (1990). *"Wissenschaft und Lebensraum," geographische Ostforschung im Nationalsozialismus: Ein Beitrag zur Disziplingeschichte der Geographie.* Hamburger Beiträge zur Wissenschaftsgeschichte 8. Berlin: Dietrich Reimer Verlag.

Rohrbach, Paul (1926). *Deutsches Volkstum als Minderheit.* Berlin: Hans Robert Engelmann.

Romer, Eugeniusz (1916). *Atlas Polski.* Warsaw, Cracow: Freytag & Berndt.

——— (1918). *Atlas Polski.* Partial and translated version of the 1916 edition. Privately published by A. Jechalski in the United States.

Rosenberg, Alfred (1936). *Gestaltung der Idee. Reden und Aufsätze 1933–35.* Blut und Ehre, ed. Thilo von Trotha, vol. 2. Munich: Zentralverlag der NSDAP.

——— (1939). *Das Parteiprogramm. Wesen, Grundsätze und Ziele der NSDAP.* Munich: Zentralverlag der NSDAP.

Rüdiger, Hermann (1935). Geographie und Deutschtumskunde. In *Verhandlungen und wissenschaftliche Abhandlungen des 25. Deutschen Geographentages zu Bad Nauheim, 22. bis 24. Mai 1934*, 85–99. Breslau: F. Hirt.

Das Saarbuch (1934). *Volk und Reich, Beiheft 3/4.*

Sack, Robert David (1986). *Human Territoriality. Its Theory and History.* Cambridge Studies in Historical Geography 7. Cambridge: Cambridge University Press.

Sämer, Eduard (1935). Die Geopolitik und ihre Behandlung in der UIrg. *Geographischer Anzeiger* 36: 522, 543–5.

Sandner, Gerhard (1983). Die "Geographische Zeitschrift" 1933–1944. Eine Dokumentation über Zensur, Selbstzensur und Anpassungsdruck bei wissenschaftlichen Zeitschriften im Dritten Reich. *Geographische Zeitschrift* 71: 65–87, 127–49.

——— (1988). Recent Advances in the History of German Geography 1918–1945: A Progress Report for the Federal Republic of Germany. *Geographische Zeitschrift* 76: 120–33.

——— (1990). Die Geographie als Wegbereiterin und als dienstbare Wissenschaft im Nationalsozialismus: Ideologische Wurzeln, fachpolitische Ziele und Wirkungen. Ringvorlesung "Wissenschaft im Faschismus." Lecture manuscript, Berlin: Projekt für interdisziplinäre Faschismusforschung, 6 June 1990.

——— (1994). In Search of Identity: German Nationalism and Geography, 1871–1910. In *Geography and National Identity*, ed. D. Hooson, 71–91. The Institute of British Geographers Special Publications Series 29. Cambridge, MA: Blackwell.

Sandner, Gerhard and Mechthild Roessler (1994). Geography and Empire in Germany, 1871–1945. In *Geography and Empire*, ed. A. Godlewska and N. Smith, 115–27. The Institute of British Geographers Special Publications Series 30. Cambridge, MA: Blackwell.

Sappok, Gerhard (1939). Deutsche Kulturleistungen in Polen. *Volk und Reich* 15: 587–602.

Schadewaldt, Hans (1939a). Der Deutschtumskampf in Westpolen. In *Deutschland und der Korridor*, ed. Friedrich Heiss, Günther Lohse, and Waldemar Wucher, 223–46. Berlin: Volk und Reich.

———— (1939b). Die polnischen Kriegsziele. *Volk und Reich* 15: 603–14.

Schäfer, Dietrich (1919). *Sprachenkarte der deutschen Ostmarken*, 1:1,000,000. Berlin: Karl Curtius.

Schäfer, Otto (1937). Politik gegen Geopolitik. Zur geopolitischen Lage der Tschechoslowakei. *Zeitschrift für Geopolitik* 14: 421–34.

Scheer, A., ed. (1927). *Fischer-Geistbeck Stufenatlas für höhere Lehranstalten. III Oberstufe.* Bielefeld, Leipzig: Velhagen & Klasing.

Schieder, Theodor (1991). *Nationalismus und Nationalstaat. Studien zum nationalen Problem im modernen Europa*, ed. Otto Dann and Hans-Ulrich Wehler. Göttingen: Vandenhoeck & Ruprecht.

———— (1992). *Das deutsche Kaiserreich von 1871 als Nationalstaat*, 2nd ed., ed. Hans-Ulrich Wehler. Göttingen: Vandenhoeck & Ruprecht.

Schmidt, Max Georg and Hermann Haack (1929). *Geopolitischer Typenatlas. Zur Einführung in die Grundbegriffe der Geopolitik.* Gotha: Justus Perthes.

Schödl, Günter (1978). *Alldeutscher Verband und die deutsche Minderheitenpolitik in Ungarn 1890–1914.* Erlanger Historische Studien 3. Frankfurt, Berne, Las Vegas: Peter Lang.

Schultz, Hans-Dietrich (1980). *Die deutschsprachige Geographie von 1800 bis 1970. Ein Beitrag zur Geschichte ihrer Methodologie.* Abhandlungen des Geographischen Instituts, Anthropogeographie 29. Berlin: Selbstverlag des Geographischen Institut der Freien Universität Berlin.

———— (1987). Pax Geographica: Räumliche Konzepte für Krieg und Frieden in der geographischen Tradition. *Geographische Zeitschrift* 75: 1–22.

———— (1989a). Deutschlands "natürliche" Grenzen. "Mittellage" und "Mitteleuropa" in der Diskussion der Geographen seit dem Beginn des 19. Jahrhunderts. *Geschichte und Gesellschaft* 15: 248–81.

———— (1989b). Fantasies of "Mitte". "Mittellage" and "Mitteleuropa" in German Geographical Discussion in the 19th and 20th Centuries. *Political Geography Quarterly* 8: 315–39.

———— (1990). Deutschlands "natürliche" Grenzen. In *Deutschlands Grenzen in der Geschichte*, ed. Alexander Demandt, 33–88. Munich: C. H. Beck.

———— (1995). Was ist des Deutschen Vaterland? *Geographische Rundschau* 47: 429–97.

Schumacher, Rupert von (1934). Zur Theorie der Raumdarstellung. *Zeitschrift für Geopolitik* 11: 635–52.

———— (1935). Zur Theorie der geopolitischen Signatur. *Zeitschrift für Geopolitik* 12: 247–65.

Schumacher, Rupert von and Hans Hummel (1937). *Vom Kriege zwischen den Kriegen. Die Politik des Völkerkampfes.* Stuttgart: Union Deutsche Verlagsgesellschaft.

Schwabe, Klaus (1971). *Deutsche Revolution und Wilson-Frieden.* Düsseldorf: Droste Verlag.

Schwalm, Eberhardt (1978). Probleme der polnischen und deutschen Geschichte auf Karten deutscher und polnischer Schulbücher und Atlanten. *Geschichte in Wissenschaft und Unterricht* 29: 128–39.

Seeba, Hinrich C. (1994). "Germany – A Literary Concept?" The Myth of National Literature. *German Studies Review* 17: 353–69.

Seton-Watson, Hugh (1977). *Nations and States. An Enquiry into the Origins of Nations and the Politics of Nationalism.* Boulder, CO: Westview Press.

Seymour, Charles (1921). The End of an Empire: Remnants of Austria-Hungary. In *What Really Happened at Paris. The Story of the Peace Conference, 1918–19*, ed. Edward Mandell House and Charles Seymour, 87–111. New York: Charles Scribner's Sons.

—————— (1928 III). *The Intimate Papers of Colonel House*. Vol. 3, *Into the War (April 1917–June 1918)*. Boston: Houghton Mifflin Co.

—————— (1960). *Geography, Justice, and Politics at the Paris Peace Conference of 1919*. Bowman Memorial Lectures, Series One. New York: The American Geographical Society, 1951; reprinted in *The Versailles Settlement. Was it Foredoomed to Failure?*, ed. Ivo Lederer, 106–16. Boston: D. C. Heath & Co. (page references are to reprint edition).

Shotwell, James T. (1937). *At the Paris Peace Conference*. New York: Macmillan.

Sieger, Robert (1919). Massgebliches und Unmassgebliches. Die künftigen Grenzen Deutschösterreichs. *Grenzboten* 78,4: 244–6.

—————— (1924). Rudolf Kjellen. *Zeitschrift für Geopolitik* 1: 339–46.

Smith, Arthur D. Howden (1918). *The Real Colonel House*. New York: George H. Doran.

Smith, Woodruff D. (1986). *The Ideological Origins of Nazi Imperialism*. New York: Oxford University Press.

Smogorzewski, Casimir (1930). *Poland, Germany and the Corridor*. London: Williams and Norgate.

Soffner, Heinz (1942). War on the Visual Front. *American Scholar* 11: 465–76.

Späne der Arbeitsgemeinschaft für Geopolitik (1936). Emigranten fälschen Karten. *Zeitschrift für Geopolitik* 13: 404–5.

Speier, Hans (1941). Magic Geography. *Social Research* 8: 310–30.

Spett, Jakob (1918). *Nationalitätenkarte der östlichen Provinzen des Deutschen Reiches nach den Ergebnissen der amtlichen Volkszählung vom Jahre 1910*, 1:500,000. Vienna: Moritz Perles.

Spohr, W. (ca. 1930). *Deutsche Brüder im Osten*. Auslanddeutsche Volkshefte 4. Berlin: Verlagsanstalt H. A. Braun.

Springenschmid, Karl (1934). *Die Staaten als Lebewesen*, 2nd ed. Leipzig: E. Wunderlich.

—————— (1935). *Deutschland und seine Nachbarn. Eine geopolitische Bildreihe*. Leipzig: E. Wunderlich.

—————— (1936). *Deutschland, geopolitisch gesehen*. Leipzig: E. Wunderlich.

Stahlberg, Walter (1920). Mußte das sein? Ein Stück vom politischen Polen und vom unpolitischen Deutschen. *Eiserne Blätter* 1,44: 767–72.

—————— (1921). Das Kartenspiel um Oberschlesien. *Grenzboten* 80,3: 6–27.

Stark, Gary D. (1981). *Entrepreneurs of Ideology. Neoconservative Publishers in Germany, 1890–1933*. Chapel Hill: University of North Carolina Press.

Stiftung für deutsche Volks- und Kulturbodenforschung (1930). Die Grenzziehung in einzelnen typischen Beispielen. In *Die Auswirkungen der Gebietsabtretungen auf die deutsche Wirtschaft*, vol. 1, *Der Osten und Norden*, ed. Ausschuß zur Untersuchung der Erzeugungs- und Absatzbedingungen der deutschen Wirtschaft, 102–12. Berlin: Ernst Siegfried Mittler & Sohn.

Stillich, Oskar (1930). *Fort mit dem VDA aus den Schulen*. Breslau: privately published.

Strausz-Hupé, Robert (1942). *Geopolitics. The Struggle for Space and Power*. New York: G. P. Putnam's Sons.

Ströhle, Albert (1931). *Von Versailles bis zur Gegenwart. Der Friedensvertrag und seine Auswirkungen*. Berlin: Zentralverlag.

Stuart, Campbell (1920). *Secrets of Crew House. The Story of a Famous Campaign.* London: Hodder & Stoughton.

Sywottek, Jutta (1976). *Mobilmachung für den totalen Krieg. Die propagandistische Vorbereitung der deutschen Bevölkerung auf den Zweiten Weltkrieg.* Studien zur modernen Geschichte 18. Opladen: Westdeutscher Verlag.

Taylor, Peter J. (1989). *Political Geography. World Economy, Nation-State and Locality,* 2nd ed. London: Longman.

Temperley, H. W. V., ed. (1969 II). *A History of the Peace Conference of Paris.* Vol. 2, *The Settlement with Germany.* London: Oxford University Press, 1920; reprint 1969.

Temperley, H. W. V., ed. (1969 III). *A History of the Peace Conference of Paris.* Vol. 3, *Chronology, Notes and Documents.* London: Oxford University Press, 1920; reprint 1969.

Thalheim, Karl C. and A. Hillen Ziegfeld (1936). *Der deutsche Osten. Seine Geschichte, sein Wesen und seine Aufgabe.* Berlin: Propyläen Verlag.

Thies, Johann (1932). Geopolitik in die Volksschule. *Zeitschrift für Geopolitik* 9: 503–12, 626–37.

Thomas, Louis B. (1949). Maps as Instruments of Propaganda. *Surveying and Mapping* 9,2: 75–81.

Tiessen, Ernst (1924). Der Friedensvertrag von Versailles und die politische Geographie. *Zeitschrift für Geopolitik* 1: 203–20.

Tolstoy, Alexei (1953). *Ordeal.* Vol. 3. Moscow: Foreign Languages Publishing House.

Trampler, Kurt (1931). Die Sendung der Deutschen im friedlosen Europa. In *Deutschlands Weg an der Zeitenwende,* ed. K. Haushofer and K. Trampler. Munich: Hugendubel.

——— (1934a). Deutsche Grenzen. *Zeitschrift für Geopolitik* 11: 15–71.

——— (1934b). *Der Unfriede von Versailles, ein Angriff auf Volk und Lebensraum.* Munich: J. F. Lehmanns Verlag.

——— (1934c). *Volk ohne Grenze. Mitteleuropa im Zeichen der Deutschenverfolgung.* Stuttgart: Verlag Grenze und Ausland.

——— (1935). *Um Volksboden und Grenze.* Schriften zur Geopolitik 9. Heidelberg: K. Vowinckel.

Tröger, Jörg, ed. (1984). *Hochschule und Wissenschaft im Dritten Reich.* Frankfurt: Campus Verlag.

Troll, Carl (1947). Die geographische Wissenschaft in Deutschland in den Jahren 1933 bis 1945: Eine Kritik und Rechtfertigung. *Erdkunde* 1: 3–48.

Tyner, Judith (1974). Persuasive Cartography: An Examination of the Map as a Subjective Tool of Communication. Ph.D. diss., University of California-Los Angeles.

——— (1982). Persuasive Cartography. *Journal of Geography* 81,4: 140–4.

Uhlig, Karl (1919). Neue Karten zur gegenwärtigen politischen Lage. *Mitteilungen des Deutschen Ausland-Instituts* 2,8: 284–6.

Ulitzka, Prälat (1932). Traurige Gedenktage. *Der Heimatdienst* 12,6: 182.

Unser Staat und der Weltfrieden (1938). *Deutsche Arbeit* 38,7: 278–81.

Verhandlungen des 20. Deutschen Geographentages zu Leipzig, vom 17. bis 19. Mai 1921 (1922). Berlin: Dietrich Reimer.

Verhandlungen des 21. Deutschen Geographentages zu Breslau, vom 2. bis 4. Juni 1925 (1926). Berlin: Dietrich Reimer.

Verhandlungen des 25. Deutschen Geographentages zu Bad Nauheim, vom 22. bis 24. Mai 1934 (1935). Breslau: F. Hirt.

Verhandlungen des 26. Deutschen Geographentages zu Jena, vom 9. bis 12. Oktober 1936 (1937). Breslau: F. Hirt.

Viererbl, Karl (1937). Memoire III. *Volk und Reich* 13: 380–401.

Voigt, Gerd (1965). Aufgaben und Funktion der Osteuropa-Studien in der Weimarer Republik. In *Studien über die deutsche Geschichtswissenschaft*, vol. 2, *Die bürgerliche deutsche Geschichtsschreibung von der Reichseinigung von oben bis zur Befreiung Deutschlands vom Faschismus*, ed. Joachim Streisand, 369–99. Berlin: Akademie Verlag.

Völkel, R. (1940). *Deutschland. Erdkundebuch für höhere Schulen*, vol. 8, ed. E. Hinrichs. Frankfurt: Moritz Diesterweg.

Die völkerrechtlichen Grenzen des Deutschen Reiches auf Grund des Selbstbestimmungsrechts der Völker (ca. 1989). Vlotho, Weser: Verlag für Volkstum und Zeitgeschichtsforschung.

Volz, Jochim (1930/1). Untersuchungen über Grenzzerreissungsschäden. *Deutsche Hefte für Volks- und Kulturbodenforschung* 1: 39–42.

Volz, Wilhelm (1920). *Zwei Jahrtausende Oberschlesien*. Breslau: Grass, Barth & Co. W. Friedrich.

—— (1921a). *The Economic-Geographical Foundations of the Upper Silesian Question*. Berlin: Georg Stilke.

—— (1921b). *Die völkische Struktur von Oberschlesien*. Breslau: M. & H. Markus.

—— (1922). Oberschlesien und die oberschlesische Frage. *Zeitschrift der Gesellschaft für Erdkunde zu Berlin*, 161–234.

—— ed. (1925). *Der westdeutsche Volksboden. Aufsätze zu Fragen des Westens*. Breslau: F. Hirt.

—— ed. (1926). *Der ostdeutsche Volksboden. Aufsätze zu Fragen des Ostens*. Breslau: F. Hirt.

—— (1942). Ein halbes Jahrhundert Geograph. Ein Rückblick von W. Volz. *Zeitschrift für Erdkunde* 10: 717–33.

Volz, Wilhelm and Hans Schwalm (1929). *Die deutsche Ostgrenze. Unterlagen zur Erfassung der Grenzzerreissungsschäden*. Langensalza, Berlin, Leipzig: Kommissionsverlag Julius Beltz.

Wagner, Carl (1938). Ansprache, 2. Tagung der Deutschen Kartographischen Gesellschaft e.V., vom 21. bis 23. Oktober 1988 in Berlin. *Mitteilungen des Reichsamts für Landesaufnahme* 14: 341–5.

Waldmann, Karl (1984). Die Geographie an der Friedrich-Alexander-Universität Erlangen 1919 bis 1945. Die Entwicklung einer Hochschuldisziplin unter verschiedenen Rahmenbedingungen. Staatsexamensarbeit, University of Erlangen.

Weber, Paul (1914). *Die Polen in Oberschlesien. Eine statistische Untersuchung*. Berlin: J. Springer.

Wehler, Hans-Ulrich (1988). *Entsorgung der Vergangenheit? Ein polemischer Essay zum "Historikerstreit"*. Munich: C. H. Beck.

Weidenfeller, Gerhard (1976). *VDA. Verein für das Deutschtum im Ausland. Allgemeiner Deutscher Schulverein (1881–1918). Ein Beitrag zur Geschichte des deutschen Nationalismus und Imperialismus im Kaiserreich*. Europäische Hochschulschriften, series III, vol. 66. Frankfurt: Peter Lang; Berne: Herbert Lang.

Weigert, Hans (1941). Maps are Weapons. *Survey Graphics* 10 (October): 528–30.

Weinstein, J. (1931). *Oberschlesien. Das Land der Gegensätze*. Paris: Gebethner & Wolff.

Weissbecker, Manfred (1983). *Bund Deutscher Osten* (BDO). In *Lexikon zur Parteiengeschichte. Die bürgerlichen und kleinbürgerlichen Parteien und Verbände in Deutschland (1789–1945)*, ed. Dieter Fricke et al., vol. 1, 308–15. Cologne: Pahl-Rugenstein; Leipzig: Bibliographisches Institut.

Weizsäcker, Wilhelm (1926). Die Ausbreitung des deutschen Rechtes in Osteuropa. In *Staat und Volkstum*, Bücher des Deutschtums 2, ed. Karl C. von Loesch and A. H. Ziegfeld, 549–67. Berlin: Deutscher Schutzbund Verlag.

Welch, David, ed. (1983). *Nazi Propaganda. The Power and the Limitations.* London: Croom Helm.

———— (1993). *The Third Reich. Politics and Propaganda.* London: Routledge.

Werner, Karl (1932). *Weichselkorridor und Ostoberschlesien. Der Weltwirtschaftliche Zusammenhang beider Probleme.* Zur Wirtschaftsgeographie des deutschen Ostens, vol. 2, ed. Walter Geisler. Breslau: M. & H. Marcus.

———— (1933). Schlesien und der Korridor. *Volk und Reich, Beiheft* 1, February.

Wertheimer, Mildred (1924). *The Pan-German League 1890–1914.* New York: privately published.

Whittlesey, Dernwet (1939). *The Earth and the State. A Study of Political Geography.* New York: Holt.

Whittlesey, Derwent, Charles C. Colby, and Richard Hartshorne (1942). *German Strategy of World Conquest.* New York: Farrar & Rinehart.

Wilkinson, H. R. (1951). *Maps and Politics. A Review of the Ethnographic Cartography of Macedonia.* Liverpool: Liverpool University Press.

Williams, C. and A. D. Smith (1983). The National Construction of Social Space. *Progress in Human Geography* 7: 502–18.

Winkler, Erwin (1927). *Statistisches Handbuch für das gesamte Deutschtum.* Berlin: Verlag Deutsche Rundschau.

Wippermann, Klaus W. (1976). *Politische Propaganda und staatsbürgerliche Bildung. Die Reichszentrale für Heimatdienst in der Weimarer Republik.* Cologne: Verlag Wissenschaft und Politik.

Wirsing, Giselher, ed. (1941). *The War in Maps, 1939–1940.* New York: German Library of Information.

Witt, Kurt (1935). *Die Teschener Frage.* Berlin: Volk und Reich.

Wittschell, Leo (1926). Tatsachen und Betrachtungen zur Geopolitik Ostpreussens. *Zeitschrift für Geopolitik* 3: 429–44.

Wolf, Armin (1978). 100 Jahre Putzger: 100 Jahre Geschichtsbild in Deutschland (1877–1977). *Geschichte in Wissenschaft und Unterricht* 11: 702–18.

———— (1991). What Can the History of Historical Atlases Teach? Some Lessons from a Century of Putzger's "Historischer Schulatlas". *Cartographica* 28,2: 21–37.

Wright, John Kirtland (1942). Map Makers are Human. Comments on the Subjective in Maps. *The Geographical Review* 32: 527–44.

———— (1952). *Geography in the Making. The American Geographical Society, 1851–1951.* New York: American Geographical Society.

Wright, Jonathan and Paul Stafford (1988). A Blueprint for World War? Hitler and the Hossbach Memorandum. *History Today* 38,3: 11–17.

Zeman, Z. A. B. (1973). *Nazi Propaganda.* New York: Oxford University Press.

———— (1982). *Selling the War: Art and Propaganda in World War II.* New York: Exeter Books.

Ziegfeld, Arnold Hillen (1925). Die deutsche Kartographie nach dem Weltkriege. In *Volk unter Völkern*, Bücher des Deutschtums 1, ed. Karl C. von Loesch and A. H. Ziegfeld, 429–45. Breslau: F. Hirt.

———— (1926). Karte und Schule. In *Staat und Volkstum*, Bücher des Deutschtums 2, ed. Karl C. von Loesch and A. H. Ziegfeld, 705–29. Berlin: Deutscher Schutzbund Verlag.

———— (1935). Kartengestaltung – Sport oder Waffe? *Zeitschrift für Geopolitik* 12: 243–47.

Ziegfeld, Arnold Hillen and Wilhelm von Kries (1933). Das Korridorproblem –
als Wirklichkeit. In *Deutschland und der Korridor*, ed. Friedrich Heiss and
A. H. Ziegfeld, 275–404. Berlin: Volk und Reich.
Zu der Luth, Rudolf (1934). *Wehrwissenschaftlicher Atlas*. Vienna: Kommissionsverlag
Josef Lenobel.
Zur Bevölkerungskarte der Ober- und Niederlausitz (1925). *Mitteilungen der Gesell-
schaft für Erdkunde zu Leipzig*, 158–61.

INDEX